About the Series

IDEAS IN PROGRESS is a commercially published series of working papers dealing with alternatives to industrial society. It is our belief that the ills and profound frustrations which have overtaken man are not merely due to industrial civilization's inadequate planning and faulty execution, but are caused by fundamental errors in our basic thinking about goals. This series is designed to question and rethink the underlying concepts of many of our institutions and to propose alternatives. Unless this is done soon society will undoubtedly create even greater injustices and inequalities than at present. It is to correct this trend that authors are invited to submit short texts of work in progress of interest not only to their colleagues but also to the general public. The series fosters direct contact between the author and the reader. It provides the author with the opportunity to give wide circulation to his draft while he is still developing an idea. It offers the reader an opportunity to participate critically in shaping this idea before it has taken on a definite form. Readers are invited to write directly to the author of the present volume at the following address:

William Leiss, Faculty of Environmental Studies, York University, 4700 Keele Street, Downsview, Ontario, Canada.

About the Author

William Leiss was born in the United States in 1939 and was educated at Fairleigh Dickinson, Brandeis and the University of California. He is now Professor in Environmental Studies and Professor of Political Science at York University, Toronto, Canada. His previous books include *The Domination of Nature* and a previous version of this book which was published in 1976. He has also edited *Environment and Politics in Canada*, a collection of essays, and has contributed numerous reports and articles in American, English, Canadian and Spanish journals. He lives in Toronto.

IDEAS IN PROGRESS

The Limits to Satisfaction: on needs and commodities

William Leiss

MARION BOYARS · LONDON

First published in Great Britain in 1978 by
Marion Boyars Publishers Ltd.
18 Brewer Street, London W1R 4AS

This is a revised edition of *The Limits to Satisfaction—an essay on the problem of needs and commodities* published in 1976 by the University of Toronto Press

Australian distribution by Thomas C. Lothian
4-12 Tattersalls Lane
Melbourne, Victoria 3000

ISBN 0 7145 2646 0 (cased edition)
ISBN 0 7145 2647 0 (paperback edition)

Printed in Great Britain by
Whitstable Litho Ltd.

Contents

for my mother and brothers, and for Miriam

Preface

In the year 1972 the 210 million people of the United States, in pursuit of the satisfaction of their needs, used approximately 4.4 billion short tons of new material resources (minerals and non-food organic materials). This averages 42,500 pounds per person - for a single year; included were 17,800 pounds of natural gas, coal, and petroleum, 17,500 pounds of stone, sand, and gravel, 2,750 pounds of forest products, 1,200 pounds of iron, 50 pounds of aluminum, 25 pounds of copper, and 15 pounds of zinc.

At the same time, the U.S. Bureau of Labor Statistics reported that one-half of American families did not have sufficient income to enjoy a 'moderate' standard of living. Yet these families lived in the nation that enjoyed a disproportionately large share of the earth's bounty which was commanded by the 'developed' countries, that fortunate one-third of the human population which consumed ninety per cent of the total resource production.

And what of the future? A U.S. government commission projected the average yearly per capita demands of 300 million U.S. citizens in the year 2000 as follows: 5,000 pounds of forest products, 1,600 pounds of iron, 200 pounds of aluminum, 67 pounds of copper, and 23 pounds of zinc. The annual energy requirements of each U.S. citizen at that time would be the equivalent of twenty metric tons of coal.

In the light of the enormity of these demands upon the earth's natural resources, is it surprising that there should be widespread public concern in the 1970s over shortages and the possible scarcity of energy and minerals? That groups of experts should consider the feasibility of extracting resources from other planets? That haggling over the right to plunder the earth on favourable terms - such as the pointed reminder in

1974 by the International Economic Policy Association that the United States requires 'continued and reliable access to Canadian natural resources' – should cast a pall over the future relationships of all nations?

Short-run solutions to the problem of resource supply probably will be found, at least for the developed countries, whose economic and political power guarantees their disproportionate share of the world's production in the immediate future. Yet so long as public discussion is concerned almost exclusively with the adequacy of resource supplies, another equally significant aspect of our situation will continue to be neglected. For while much attention has been focused on problems of supply, far too little has been bestowed upon problems of demands, wants, and needs.

At its roots the problem of material demands in modern industrialized societies is nothing less than the problem of the possibilities of human satisfaction. The problem we must face now is to assess both the nature of human wants and the nature of the means by which we attempt to satisfy them. The problem of supply – that is, the provisioning of material resources for the satisfaction of needs – has no solution of itself, apart from its intrinsic relationship to the processes of social interaction wherein needs and wants arise.

In this essay I have examined the problem of human needs in the high-intensity market setting of contemporary industrialized societies. This setting promotes a lifestyle that is dependent upon an endlessly rising level of consumption of material goods. I have tried to show that, within the social process which enshrines this lifestyle as the highest goal of human endeavour, individuals are led to misinterpret the nature of their needs and to misunderstand the relationship between their needs and the ways in which they may be satisfied. And I have argued that, so long as that process persists, individuals will experience the frustration of their expectations and the terrors of anticipated scarcity in proportion to the massive, ever-growing accumulation of material wealth.

There is no prospect for a solution to this fundamental paradox apart from a better understanding of our needs and a reorientation of our social and economic behaviour away from dependence on a high-consumption lifestyle. The alternative prospect is increasingly bitter competition over available resources both within and among the industrialized nations.

W.L.
Toronto
November 1975

THE LIMITS TO SATISFACTION

Introduction

What the so-called developed world has made out of the fantastic resources at its disposal is hardly worthy of emulation by the remainder. Yet one must marvel simply at the ability of a single society to use 4.4 billion tons of new natural materials in one year.[1] This accomplishment, if we may call it that, testifies to the extraordinary collective ingenuity of the users, whatever may be the specific objectives upon which that ingenuity has been exercised, and we may pay tribute to it without belittling the equally admirable efforts of earlier human groups, who with far cruder techniques erected the monumental burial chambers of their rulers and the temples of their gods.

Civilizations have a propensity for enshrining their predominant social concerns in massive physical structures, the relative dimensions of which provide a clue to the differential urgency of the things most ardently desired. Pyramids and palaces, cathedrals and castles have thus given way to the severe glass-and-steel towers that form the horizons of cities and assert the predominance of economic affairs in everyday life. The physical landscape stands as a symbol for other differences. To a greater or lesser extent in all pre-modern societies the economy – considered as the organized production of the material necessities of life – was embedded within a total cultural milieu which regulated the circulation of goods according to social norms and determined what kinds of things could be exchanged (land and labour are not exchanged in the market in most 'primitive' societies, for example).[2] The division of labour, investment in technological innovation, and the exploitation of natural resources were governed only in part by principles of productive efficiency, for these economic activities were inseparable from other concerns, such as the maintenance of kinship, caste, and rank orders, upon which were grounded social stability and authority.

Modern society represents the first large-scale attempt to found stability and authority not upon the earlier patterns of inherited privilege or traditional associations, but rather directly on the achievements of economic production and the satisfaction of needs. Naturally this decisive change did not happen all at once; and it is a process that, after many years of development, is not yet completed. This is the tendency which the emergence of capitalism during the early modern era set in motion, however, and it may be expressed in the following way: the primary social bond is the identification of the self-interest of the individual, intent upon maximizing the satisfaction of his needs, with the interest of society as a whole, which is to maximize total productive output. In concrete terms, the well-being of every individual is thought to be identical with the steady rise of the Gross National Product.

Not everyone benefits equally, to be sure, and some may be ruined utterly in the endeavour. All that is necessary to sustain the legitimacy of this process in the contemporary period is for the majority of individuals at any time to believe that, on balance, their private interests are best served by the uninterrupted growth of the total social product. The principle of legitimacy for modern society, which simply stated is the rationale for the acceptance of the prevailing distribution of rewards and power, now consists in a permanently rising level of consumption. This principle is today at work not only in government-managed capitalist societies (North America and Western Europe), but also in the industrialized socialist nations (Eastern Europe and the Soviet Union).

The Index of Consumer Sentiment, designed to measure the fluctuating level of popular confidence in the prospects for the national economy at regular quarterly intervals, may be the most reliable barometer of social consciousness at present.[3] In any case the point of departure for this essay is my assumption that a permanently rising level of consumption is the single most important element in the network of popular hopes and fears whereby the citizens of industrialized capitalist and socialist societies register their belief in the legitimacy of those social systems.

A permanently rising level of consumption requires an adequate supply of energy and material resources. Some time ago *The Limits to Growth* created a minor public furor, among other things for introducing into the 'world model' the factor of dwindling non-renewable natural resources. The counter-attacks which were addressed to this point stressed the reasonableness of anticipating the discovery of substantial new reserves in many areas, including the vast quantities of mineral wealth potentially recoverable from sea-water.[4] What is clear from the debate is that we cannot now estimate with much assurance the magnitude of possible shortages in natural resources, although it is certain that future prices of raw materials will rise in most cases.

Projections are highly unreliable, as everyone knows. Yet while they may not be very useful for policy calculations, they do give us some perspective on the escalation of our dependence on the utilization of energy and resources. For these purposes an earlier study, published in 1963 by a U.S. research group called Resources for the Future, is illuminating. This group attempted to project cumulative demand for various minerals over the period 1960-2000 and to compare this with the known world reserves of those minerals in 1960. Forecasting world demand was found to be difficult owing to a lack of reliable information for the 'communist' nations. However, the forecasts for the 'non-communist' part of the world, where data on rates of growth were readily available, prove sufficiently interesting of themselves. In the case of nickel, tungsten, copper, lead, and zinc, the projected demand of these nations alone exceeded the known world resources![5] Undoubtedly new sources of supply could be found and developed; but should we submerge all our perspectives in purely technological considerations? Should we not also ask why certain societies embarked upon a path of social development that would necessitate sustained exploration and technological innovation simply to prevent themselves from exhausting the entire proven earthly supply of some minerals in less than two generations?

Aggregate figures, however, do not tell the most interesting part of the story. Of the many inequities in the proportionate shares of resource consumption, let us consider three. The industrially developed world, with one-third of the total human population, takes 90% of all resource production. The United States, with 5% of world population, uses 27% of all materials extracted. And some idea of the prevailing inequity within the U.S. population is conveyed by the fact that the top 20% own 76% of the total wealth, including 96% of the corporate stock.[6]

The social injustice represented by these enormous disparities in wealth has been denounced often and eloquently, to little avail. Let us then pursue another line of inquiry. Let us assume that our objective is to achieve a rough equality in the distribution of wealth (or at least much more fairness than now exists) and that the objective appears attainable. On this assumption, how much would suffice for our society as a whole?

To entertain this question we might have to enter that nether realm of metaphysical speculation which is often considered to be off-limits for responsible intellects. For so long as the way is open for enterprising talents to reap the benefits of cleverness and industry regardless of the circumstances of their birth, and for those less motivated and less richly endowed by nature to be lifted steadily by the rising base level of gratification, could we not safely ignore any questions about the steady escalation of our material demands? If the general hope for ever greater numbers of ever more attractive goods can be sustained, it is inconsequential that - apparently - so little enduring satisfaction should be provided by those now available. If we assume that every plateau of satiation merely affords a glimpse of vistas still more magnificent, and that the ascent has no natural terminus, all that remains is to organize the hunt for sufficient materials with which to construct the landscape of desire. But two major obstacles intervene.

In the first place the matter of distributive justice is not so easily resolved. Recall the 'official' projection for U.S. demand in the year 2000. The 300 million inhabitants of that nation will require for that year alone 240 million short tons of iron, 26.4 million of aluminum, 10 million of copper, and 3.5 of zinc; 22.3 billion cubic feet of forest products; and energy equivalent to 6 billion metric tons of coal. Supposing that these consumers have reached an amicable understanding among themselves for sharing out this cornucopia, what will be the lot of their relations dwelling elsewhere on the globe, who are expected to number seven billion by that time? What prodigies of technological ingenuity and social organization would be required were all of them to stem the burgeoning tide of their wants in the same manner? And if the great majority are not to be permitted to do so, what devices of force and cajolery will enable those 300 million to revel undisturbed in this immense heap of 'goods'?

Yet this first dilemma pales into insignificance beside the second. For we must face the real possibility that, even if this bounty were forthcoming, and even if it were distributed with some fairness, the meaning of what we call the satisfaction of needs will become increasingly ambiguous as the hapless consumer pursues the will-o'-the-wisp of gratification through the jungle of commodities. There is no apparent end to the escalation of demand and no assurance that a sense of contentment or well-being will be found in the higher reaches of material abundance. The society which promotes the ideal of the high-consumption lifestyle seems to lack any reliable measure of the improvements in the quality of life that we should expect to result from its expanding productive capacity. The personal objectives sought in the frenetic activity of the marketplace are more and more obscure.

Massive capital investments and technological innovations are planned to ensure the supply of resources and energy that we think we need. More thought ought to be given to the question as to whether this is the best approach to the problem of our needs. In our society individuals are encouraged to misinterpret the character of their needs and desires, and as a result the character of their attempts to assuage them becomes more and more ambiguous. The social whole composed of such individuals progressively rationalizes its productive capacity and simultaneously becomes more irrational in disposing of its effects.

What is chiefly lacking in our current understanding of our needs is a general framework for evaluating the damages that individuals inflict upon themselves, upon the bonds of social interaction, and upon the world of non-human nature in a society intent upon maximizing the production of commodities for the satisfaction of needs. This essay seeks to outline one possible design for that framework.

The central theme of this essay is that understanding the interaction of needs and commodities provides a way of unravelling the problem of endlessly expanding human needs in a high-intensity market setting. Commodities are material things and services offered for sale in the marketplace and purchased presumably because they provide some kind of satisfaction for an individual's wants, needs, or desires. What I call a high-intensity market setting is simply a market economy in which there is a very large number of commodities available to large numbers of people, and in which many commodities are the result of highly complex industrial production processes involving sophisticated scientific and technological knowledge. In those societies where such a market economy already exists or is in process of formation, a principal article of faith is that the economy should continue to expand and to offer an ever-widening array of commodities for consumers. And a principal concern for such societies is to ensure that a sufficient quantity of energy and material resources will be available for this purpose.

A market society of this kind exists in the government-managed capitalist societies of North America, Western Europe, and Japan. Obviously there are important differences in the productive capacity of the market economy among these nations and even within geographical regions of the individual countries. But the given description - namely that large numbers of their populations have access to a wide range of commodities - is applicable to all of them. The social élites of many non-industrialized countries also have access to such goods, but the majority of their populations do not, and the scope of their everyday market exchanges is relatively limited in comparison to the others. In the societies that have a high-intensity market economy the ability of the social system as a whole to ensure steady growth in the quantity and variety of commodities is an important factor in social stability.

At present the quantity and quality of available commodities is far lower in those industrialized nations that are nominally socialist - the Soviet Union and the countries of Eastern Europe. According to André Gorz, however, the trend of social development in those societies is similar to that in what I have called managed capitalist societies. Gorz maintains that the objective of the ruling bureaucratic apparatus in these socialist nations is 'to increase individual consumption *in order* to maintain and consolidate its own power' and to deflect or weaken any potential political challenge to its authority.[7] If Gorz is correct (and I think that he is), then the steering of popular hopes toward an expanding marketplace of goods is an important factor in the political stability of the existing régimes in those social systems as well. Although the present array of commodities is meagre by our standards, there is a structural similarity between the two systems in the emphasis on consumption as the chief route to individual satisfaction. These 'managed socialist' societies seem to have accepted the high-intensity market setting, which originated in modern capitalism, as a goal to be attained by another means.

Besides the central theme already mentioned, this essay discusses two ancillary themes as well.

One is an attempt to show the inadequacy of dealing with the character of human needs without explicit reference to the actual, everyday modes of satisfying needs in particular social systems. (The high-intensity market setting represents one mode of need-satisfaction.) Most contemporary theories of needs, especially those that draw upon a conception of 'basic human needs,' are deficient in this respect. In these theories the discussion of needs revolves around very abstract principles, such as the distinction between physiological and psychological dimensions, and rarely focuses on the concrete socio-economic reality wherein particular needs are formulated and wherein all attempts at individual satisfaction occur. I have tried to demonstrate that the problem of needs can be clarified by examining them in their dynamic interplay with specific, socially organized forms of satisfaction.

The second ancillary theme is my attempt to set the contemporary discussion of human needs in an ecological context. Just as most current theories ignore the concrete forms of need-satisfaction, so also they ignore the extra-human context of all our needs - the vast, infinitely complex 'support systems' of nature. The pattern of needs in each human social setting has an impact on the interactions among many species and on natural ecosystems as a whole, and the force of this impact increases dramatically with the coming of industrial production. Recently the environmental impact of human activities, as it affects our own future well-being and the functioning of ecosystems, has become a matter for public concern and scientific research. But this new concern has not yet made itself felt in a systematic way in approaches to the understanding of human needs. This is a difficult problem. I have not endeavoured to give a definitive account of it, but only to offer a few preliminary suggestions as to how we might proceed in incorporating a discussion of human needs within this larger perspective.

The critique of the expanding market economy and its impact on the lives of individuals was developed in the classic works of authors such as John Stuart Mill, Marx, Veblen, Durkheim, Lukács, and Tawney, and has been extended and elaborated by more recent writers.[8] A hard look at the religion of economic growth, the ideology of development as a model for all nations, and the possible adverse consequences of expanding industrial production and pollution on a global scale has been undertaken with greater frequency in the past few years.[9] Finally, several authors working both in the natural and the social sciences have insisted upon the necessity for developing a broad ecological perspective on the interrelationship of the political, economic, social, and environmental elements involved in the present directions of social change.[10] Some of these general analyses, together with a number of more specific studies that are referred to elsewhere, have proved helpful in formulating the argument of this essay; however, I have not indicated my agreement or disagreement with them on particular points.

Part One of this essay explores some of the difficulties involved in attempting to isolate the critical features of the social process in which we live that are responsible for the increasing confusion about our needs. Some of these difficulties arise from the ways in which we commonly use words and phrases such as *scarcity, the insatiability of wants*, and *the conquest of nature.* Others concern the everyday dilemmas we experience as consumers in attempting to determine and satisfy our needs. These two levels of confusion about our needs must be penetrated in order that the workings of the basic, underlying dynamic – the systematic orientation of all needs toward the realm of commodities – can be displayed.

Part Two offers a theory of the interaction between needs and commodities in the high-intensity market setting. The progressive fragmentation of individual needs into smaller and smaller elements is related to the progressive fragmentation of commodities (as the objects by which we attempt to satisfy our needs) into assortments of attributes and messages. The reciprocal interplay between both processes introduces a double ambiguity into the activity of consumption and explains why we experience increasing confusion about our needs and about our sense of satisfaction and well-being.

Part Three suggests some ways of reconsidering our needs that might assist us in overcoming the potentially dangerous tendencies of our present practices and in breaking out of the vicious circle of expanding productivity and expanding wants.

Introduction to the British Edition

Since this manuscript was completed a number of important related works have been published. These include Robert Heilbroner, *Business Civilization in Decline*; Stuart Ewen, *Captains of Consciousness: Advertising and the Social Roots of the Consumer Culture*; Tibor Scitovsky, *The Joyless Economy: An Inquiry into Human Satisfaction and Consumer Dissatisfaction*; Fred Hirsch, *Social Limits to Growth*; and Marshall Sahlins, *Culture and Practical Reason.* Some of what they have to say has a direct bearing on the issues examined in this book. I will not try to give an adequate account of these books here; but a brief discussion of certain themes in them will help to set the stage for presenting the specific approach I have chosen to use in the following chapters.

In his most recent book Robert Heilbroner gives us in very broad outline the most sensible prognosis offered to date on our society's future. He argues that the economic system of corporate 'private enterprise' will be subjected increasingly to government regulation. This does not mean that private property or structural inequalities in wealth will be abolished in the foreseeable future; but the mixed economy of public and private interests will be directed more and more by civil servants whose actions will reflect the outcome of political processes. Of equal interest is Heilbroner's reflections on the changing attitudes of individuals. There is a growing general realization, he maintains, that 'economic success does not guarantee social harmony'. In the long run – over the next century – more and more persons will experience the 'hollowness at the center of a business civilization'. This hollowness results from the universal substitution of pecuniary, impersonal exchanges for interpersonal, nonpecuniary relationships, and from the degradation of labour into nothing but a means of securing income, without intrinsic satisfaction. When hopes for continual economic growth can no longer be sustained, it will become clear that these twin developments have robbed everyday life of much of its pleasure, and individuals will turn away from them toward other, as yet uncertain, forms of social organization.*

Stuart Ewen's book recounts part of the history of a decisive turn in the course of capitalist development: the realization that 'ordinary' men and women could do a double duty for the economy, first by toiling (that much was known and appreciated) and second by consuming. Consumers are not among nature's evolutionary stunts; they must be created. The industrial work-force had been assembled by the brutal disruption of long-established ways of life, and had been kept at their appointed tasks mainly by police terror. To the extent that any training in popular ideology had been found to be desirable for them, an ascetic production ethic was supplied, for example by Henry Ford. There is much irony in the fact that the mass-production techniques implemented by Ford and others signalled the obsolescence of this ethic and the necessity for its replacement.

The ideology of consumerism took shape in the 1920s. Led by business spokesmen such as Boston's Edward Filene, from whose remarkably frank speeches Ewen draws numerous quotations, its champions detailed the ways in which the 'masses' (their term) would be permitted to enjoy a share of what they had produced. Some time would pass before this reorientation of business thinking would take hold widely, and in any case the Depression and Second World War delayed its implementation until after 1945.

But the fabrication of a consumer ideology was underway. As described by Ewen, the following themes were explicitly presented in trade journals, speeches, and books of that period: (1) increasing access to consumer goods should weaken demands for workers' control in industry (i.e. 'bolshevism'); (2) as a corollary to the first, 'participation' was redefined as the exercise of consumer choice, which was to be the new and appropriate form of 'democracy'; (3) non-English ethnic minorities would be 'Americanized' and a uniform national 'taste' created through advertising messages; (4) the spheres of production and consumption would be kept separate – for example, workplace scenes would not be used in advertising (a practice that is still quite noticeable in advertisements today); (5) the family would be opened up increasingly to social pres-

* Robert Heilbroner, *Business Civilization in Decline* (London: Marion Boyars, 1976), pp. 44, 113-7.

sures, for example by new consumption models learned in public school 'home economics' courses and brought home by the kiddies; (6) male and female roles would be manipulated, with chief attention being given to presenting women as sexual objects; (7) the realm of personal consumption skills would be transformed: individuals (especially women) would be pressured into buying manufactured items (e.g. food and clothing) rather than making them at home.

In *The Joyless Economy* Tibor Scitovsky set out to undermine the economic theory of rational behaviour and consumer sovereignty. This is made up of the following propositions: (1) what the consumer chooses to do is an accurate reflection of his tastes, i.e. his behaviour is revealed by his preferences and *vice versa*; (2) the consumer develops his own tastes and preferences independently of those of other consumers; (3) without sufficient means to satisfy all of his desires, the consumer must 'keep unsatisfied margins on all his needs and desires', in order to ensure 'any extra dollar he spends on one thing yields him as much satisfaction as that extra dollar would if he spent it on any other thing'.

Scitovsky challenges his fellow economists by turning against them their most cherished value: he claims that the theory is *unscientific* in its portrayal of human psychology. He points out that the accepted theory simply cannot account at all for the obvious fact that the individual's preferences change over time – or, more precisely, that it cannot show *why* or *how* preferences change, as they obviously do. These changes are understandable only in relation to a social process of interpersonal relationships. Moreover, there is a reciprocal relation between changing preferences and changes in the sense of satisfaction derived from any particular activity; in Scitovsky's words, the dominant paradigm 'overlooks the possibility that the same influences that modify our tastes might also modify our ability to derive satisfaction from the things that cater to our tastes'.

Expressed in its simplest terms, Scitovsky's analysis attempts to explain why the expected correlation between greater happiness and rises in real income does not occur. He refers to empirical studies done in the U.S. between 1946 and 1970: 'Over this period, almost twenty-five years, per capita real income rose by 62%, yet the proportion of people who consider themselves very happy, fairly happy, and not too happy has hardly changed at all. Our economic welfare is forever rising, but we are no happier as a result.' He claims that the conventional economic paradigms cannot account for this, and he attempts to construct a theory that will do so.

There are four dimensions. First, there is empirical evidence that people derive satisfaction from status itself, that is, from relative social ranking or interpersonal comparisons that occur at *any* income level. Second, satisfaction is derived from work, but again largely as a function of the *relative* income and 'prestige' attributes of a particular job in the social hierarchy. Third, satisfaction correlates positively with novelty in one's experiences, but our own culture tends to standardize experience and progressively reduce novelty. Fourth, material progress is translated primarily into increasing comfort. Comfort, however, is like addiction: we become accustomed to it and soon take it for granted; the presence of new comforts cease to give pleasure in themselves, and only being deprived of them makes us feel the pleasure of having them (central heating, indoor toilets, adequate quantities of food, for example). 'Taken together,' Scitovsky concludes, 'they well explain why happiness should depend so much on one's ranking in society and so little on the absolute level of one's income.'*

The background context that makes sense of this 'rank-happiness' is, of course, the market or commodity-oriented society. Most individual activity is directed at increasing income that serves as the access to purchasing goods and services. This context steadily depreciates the value of activities that do not serve this objective, including the intrinsic satisfactions that might otherwise be derived from work creativity or informal interpersonal relationships. No stable or permanent sense of satisfaction is achieved simply by virtue of the fact that most individuals have a higher 'standard of living' than preceding generations had.

Fred Hirsch's *Social Limits to Growth* complements Scitovsky's analysis. The basic similarity in the two books has to do with the social impact of changes in individual preferences. Hirsch argues that, once 'basic material necessities' are met for the majority of the population in a market-exchange economy, there is intensified competition for what he calls 'positional' goods. Since these goods are scarce by their very nature, the intensified competition for them yields no net benefits for individuals (the proportion of individuals in the population who get them remains roughly constant), and at the same time there is a high social cost incurred.

What are positional goods? These are goods which define social status differences among individuals; their value lies chiefly in the fact that some persons possess them and others do not. For example, to be able

* Tibor Scitovsky, *The Joyless Economy* (New York: Oxford University Press, 1976), pp. vii, 7, 64-5, 134-9.

to move to a suburban community to escape a decaying urban environment can be such a good; those who can do so *first* enjoy easy access to both city and countryside. But when and if many people become suburbanites the advantages are largely negated: the urban culture has declined so much that it is no longer worth visiting, and the nearby countryside has been swallowed up by the newer suburban developments. All that remains for most is the well-known emptiness of suburbia itself. Another example is access to scenic beauty or foreign cultures. When only a few persons have the means of access, the advantages to them are enormous, as is the social 'distance' between them and those who cannot afford the costs. An increasing general level of affluence opens the gates, but at the price of sharply devaluing the experience, due to overcrowding and the resultant deterioration in the quality of the sites.

The deterioration of overcrowded sites is an example of the social costs resulting from positional competition. A better example can be drawn from the relationship between jobs and educational requirements. It is a well-known fact that the level of educational qualifications for jobs has been steadily increasing, and that in most cases this bears little or no relation to the requisite job skills. The flaw lies in individual assumptions about the expected correlation between formal education and high-paying jobs. The proportion of 'top' positions in society remains approximately the same, but larger numbers of individuals now compete for them. Stiffer educational qualifications are one of the screening mechanisms used to sort out the competitors. The same proportion succeeds now as in the past; but society pays the enormous cost of larger facilities for formal education for all the competitors.

In different ways both Scitovsky and Hirsch are concerned with one of the key aspects of a competitive, market-oriented society which has reached a certain general level of material affluence: the importance of the symbolic attributes of goods, and the ways in which rank and status are attached to them. This is by its very nature an intractable problem within the self-imposed limits of that society. When *relative position* is at stake, then the society will and must create new securities – that is, new *symbols of success* to be striven for – at every turn. It matters little what is chosen to signify status differences. The important point is that there is no limit to the process.

Hirsch understands what he calls positional competition as intrinsically related to a commodity-oriented economy: one of his chapters is entitled 'The New Commodity Fetishism'. The positional economy is in fact largely an expression of a 'bias to material commodities' and a 'commercialization effect'; this means that individuals increasingly interpret their well-being as the degree of success they have in 'turning over' their possessions, i.e. in having what the latest fashion dictates.

Hirsch uses the phrase 'commodity fetishism' in a book which, while it is highly critical of our present society, is not 'Marxist' in the usual sense. Yet of course this concept is one of the great hallmarks in the Marxist critique of capitalism. Most authors who write in the Marxist tradition continue to use it as a concept which to them accurately depicts certain key aspects of capitalist society down to the present day. On further examination, however, there appear to be serious difficulties in this approach. Marx used this term in a very restrictive sense, and used in this sense it does not have very wide relevance for a critical assessment of contemporary society. Moreover, there is a basic flaw in Marx's conception that has gone largely unnoticed since his time.

Marx developed the notion of commodity fetishism from his discussion of use-value and exchange-value. The former constitutes the 'matter' of the commodity, the latter its 'form'. The crucial aspect of this analysis is that the fetishism of which he speaks *arises only with respect to the form of the product.* Marx refers to the 'mystical character of the commodity' which arises out of its dual nature: commodities are 'sensuous things which are at the same time supra-sensible or social'. He also states: 'The mysterious character of the commodity-form consists therefore simply in the fact that the commodity reflects the social characteristics of men's own labour as objective characteristics of the products of labour themselves, as the socio-natural properties of these things. . . .'

These passages are so familiar, so much taken for granted after repeated citation, that we rarely ask the obvious questions: What exactly is the 'mystery' that is alluded to here? The division of labour has stamped the products of human activity with a social character as far back as our anthropological researches permit us to go. Was not the dual character of objects – as sensuous things and as objects whose significance is established by cultural forms – always quite obvious? On the basis of what evidence does Marx assume that people actually are mystified by this duality? Or that people do not recognize the characteristics of their labour in the properties of produced objects?

Marx offers an analogy between the commodity and the gods of religious belief that makes clear what is the essential point about the commodity: it is a physical object that *appears* to have a 'life of its own'. This is consistent with the understanding of fetishes in 'primitive' religion: a fetish is an object which itself is thought to possess certain powers, thus differing from an idol, which only *symbolizes* the power of, say, a deity.

Three points are worthy of note. First, Marx states that the commodity form has '*absolutely* no connection with the physical nature of the commodity' (my emphasis). Thus it is not something arising out of the interplay of form and matter in the commodity – that is, the inherent duality of the commodity itself – that produces the mystery; its mysterious character is solely a function of the form alone. Second, *all* systems of goods-exchange which are sufficiently extensive to require a separate commodity as a medium of exchange have this result: the fetishism of commodities bears no intrinsic relation to the capitalist mode of production. Third, Marx gives no indication *how* the fetishism occurs under capitalist relations. In other words, if the commodity *form* is the generalized fetish, what specific kinds of fetishistic activities occur?

In order to develop his point, Marx contrasts the medieval and modern periods. Economic relations in the former are largely 'services in kind and payments in kind', and 'the social relations between individuals in the performance of their labour appear at all events as their own personal relations, and are not disguised as social relations between things, between the products of labour'. One might ask first whether this is an accurate picture of medieval social relations, i.e. whether labour did 'appear' as a form of *personal* relation, or whether the fundamental distinction between noble and non-noble (which may have been a kind of fetishism also) was not in fact the different disguise of that period.

There may be little disagreement with the implicit suggestion made here, namely that there are different forms of reification in social relations. The question remains: What exactly is the reification in commodity production? What exactly is 'the whole mystery of commodities, all the magic and necromancy that surrounds the products of labour on the basis of commodity production'? At the end of his chapter on commodity fetishism, Marx refers to what he regards as a series of conceptual errors by earlier economic theorists: these are apparently the source of the fetishism. It seems somewhat of an exaggeration to speak of them as 'magic' and 'necromancy', but this may be merely quibbling. The important point is that the kinds of notions Marx refers to were gradually rejected in the further development of 'bourgeois' economic theory – in other words, they reflect the immature phase of a discipline which was attempting the difficult task of representing the new and complex mechanisms of a generalized market exchange economy.*

* Marx, *Capital,* vol. I, tr. Ben Fowkes (Middlesex, Eng: Penguin Books, 1976), pp. 163-77 *passim.*

I would like to pose a series of questions: (1) Did Marx's concept of commodity fetishism refer only (or chiefly) to 'ideological' elements in economic theories up to his day? (2) If this concept is still useful today, what specific notions in contemporary non-Marxist economic theories are expressions of commodity fetishism? (3) Did Marx mean that 'ordinary individuals' in the capitalist society of his day, as opposed to economists, were mystified – i.e. made mistakes in their choices or opinions – because of the way the system of commodity production operated? If so, what specific kinds of mistakes did they make? (4) Is it the case that individuals today make the same (or different) mistakes for the same reasons?

One can give a clear affirmative answer only to the first of these questions. If this is the case, then the concept of commodity fetishism has a narrow range of application. Moreover, as Marshall Sahlins has argued, there is a crucial flaw in Marx's approach that weakens its critical thrust. Marx assumed that both needs and utilities were 'objective' data that could be specified without ambiguity. This is erroneous. Sahlins notes that for Marx the commodity as a use-value 'is perfectly intelligible: it satisfies human needs'. He refers to the passages in which Marx states that there is nothing mysterious in the properties of objects in so far as they are use-values or utilities, and he comments: '. . . all utilities are symbolic. Insofar as "utility" is the concept of "need" appropriate to a certain cultural order, it must include a representation, by way of concrete properties of the object, of the differential relations between persons – as contrasts of colour, line, or fabric between women's clothes and men's signify the cultural valuation of the sexes. The "system of needs" must always be relative, not accountable as such by physical necessity, hence symbolic by definition.'* The idea of the *symbolic constitution of utility* is indispensable for a critique of consumer behaviour in an industrially-advanced society.

If the commodity *qua* commodity has an enigmatic character, i.e. if it has such a character solely by virtue of its form, then one of two conclusions must follow. Either there is no problem here, as the apologists for market society claim; or, if there is a problem (as its critics say), there is no solution to it. For no industrial society, however dedicated to the ideals of communism, could abolish the commodity form entirely, at least not without erecting an absurd dictatorship of tastes in its place. If there is only direct production for use in small community groups,

* Marshall Sahlins, *Culture and Practical Reason* (Chicago: University of Chicago Press, 1976), p. 150.

then each group can suit its peculiar tastes; but this restricts the assortment of goods to what handicrafts can produce. Industrial production depends on extensive exchanges; the expression of individual preferences and a pricing mechanism – perhaps limited in its scope by express policies on the basis of welfare considerations – has a place in facilitating those exchanges.

But we cannot begin to understand how to design such policies – that is, how to limit the destructive effects of commodity-oriented consumption patterns – if we do not recognize the symbolic constitution of utility. It is the key to Scitovsky's dilemmas of rank-happiness, to Hirsch's dilemmas of the positional economy – and to the fetish of the commodity (*not* the commodity-form) as *the embodiment of psychologically-grounded attributes*. (For example, the association of automobiles with animals and the qualities conventionally ascribed to them.)

I think we must agree with Sahlins that all utilities in all cultures are symbolic. In a society like ours, where large numbers of people participate daily in extensive market exchanges, there is a *double* symbolic process at work. One facet of it is the symbolism consciously employed in the manufacture and sale of the product, including the imagery employed in the advertising designs. The second facet is the symbolic associations selectively employed by consumers in 'constructing' lifestyle models; the whole marketplace is divided into semi-autonomous sectors which respond to different cues or to the same cues in different ways.

The analytical model developed in Part Two of this book is an attempt to sketch the main features of this double symbolic process. The dialectic of needs and commodities outlined there is intended to provide one comprehensive explanation of the social processes that Ewen, Scitovsky, Hirsch, Sahlins – and Marx, at a much earlier stage – have also sought to understand from a variety of different perspectives.

The outstanding dissimilarity between my approach and the others' is the great importance I attach to one special component of these issues: the relation between humanity and non-human nature. In the daily concern with economic indicators in industrially-advanced societies there is now no place for this component. Yet we cannot continue indefinitely to ignore the effects of our impact on the regenerative capacity of the biosphere. Our indifference to it up to this point, I have tried to show, mirrors our growing indifference to the qualities of our own wants and needs. Without a better idea of how these two processes are related we will be unable to understand why the contentment that industrial progress was supposed to bring continues to elude us.

PART ONE: EXAMINATION

The individual

In order to examine the problem of human needs in the high-intensity market setting, let us begin at the concrete level of everyday experience. In an expanding market economy the individual is confronted by an increasing variety of goods and services, all of which are supposed to have some relation to his needs or desires. Obviously he must make selections among them according to his means, and in doing so he interprets his needs in relation to the range of available commodities that may satisfy them. What problems arise in this situation?

I suggest that we should approach this question naïvely, so to speak, without making any assumptions about how an individual is supposed to behave rationally or efficiently in the role of consumer. In other words, we should assume only that an individual is faced with the problem of continually reinterpreting his needs in the context of a rapidly changing array of goods and services. After looking at the nature of this problem we can then proceed to a more analytical level and ask what it means to say - as is commonly done - that an individual's wants are insatiable.

THE INDIVIDUAL AS CONSUMER

'When placed in the mouth, the [Schwerin Salivation] tester indicates amount of salivary flow on a small bobbin of tightly wound paper impregnated with a nontoxic and tasteless dye that is sensitive to saliva. The saliva seeps through the paper windings, producing dye spots which can be counted to give an indication of salivary flow, which is assumed to be correlated with appetite appeal.'[1]

The age-old conceit about the radical differences between man and beast dies hard. Judging by the above quotation, one could say that researchers in the field of consumer psychology are putting it to rest. For them the discovery of the creature known as a consumer was an event as significant as the ornithologist's sighting of a new species or the anthropologist's encounter with a hitherto-unknown primitive tribe. Few other biological specimens detected in the same period have had greater material and intellectual resources lavished on the study of their behaviour. Vast research programs undertaken by both universities and private firms have sought to analyze and predict the vagaries of this creature in its natural setting, the 'living laboratory of the marketplace.'[2]

One of the leading authorities in this field has remarked: 'Consumers constantly need new stimulation; influencing them over the short run may not assure that they remain so stimulated over extensive periods.'[3] In the oppressive ancient empires the aristocracy's retinue were set to work devising novel ways of reinvigorating the jaded appetites of their masters; one can see in sources such as Petronius' *Satyricon* how much ingenuity was required for embellishing nature's bounty intricately enough to avoid offending decadent tastes. Clearly the ancient productive system could allow very few the luxury of tuning their desires so exquisitely. However, modern technology and industrial productivity seem to have defeated this restriction, thereby allowing everyone to modulate his or her wants according to the changing notes sounded by the pitch-pipe of the market.

Certainly the number of choices open to so many individuals, as they seek the satisfaction of their needs in the expanding marketplace, bears no comparison with earlier historical periods. However, a careful look at this situation reveals some significant and deeply rooted problems, among which are the following: the inadequacy of the individual's knowledge and 'common-sense' judgments with respect to the composition of goods; the danger of physiological and psychological damage in the use of products; the deteriorating quality of decisions owing to the scarcity of time; and the confused objectives of the person's wants themselves.

Every individual is faced with the problem of 'matching' his needs to the opportunities that are available for satisfying them in any social setting. In the high-intensity market setting individuals are encouraged to orient their needs toward the kinds of satisfactions that are embodied in the expanding array of commodities.[4] In everyday life, therefore, the individual must be able to judge the suitability of various commodities in relation to his perceptions of his needs. If these judgments are not to be purely arbitrary, the individual must have some knowledge about the qualities or characteristics of the things by means of which his needs are to be satisfied. Yet any attempt to obtain adequate knowledge for this purpose at present would appear to be fraught with great difficulties.

In the first place there is the sheer number and variety of products that become available in this market setting from year to year. Considering only those items intended for sale in supermarkets, we find that each year in North America 1,500 new products appear, 80% of which are withdrawn within that year to be replaced by another assortment.[5] How is it possible to imagine that individual decisions about the suitability of so many things for our needs could be based upon knowledge about their qualities? The kind of knowledge that seems appropriate for judging the suitability of things in relation to the objectives of our needs is the knowledge that is applied in craft skills. Examples are the cook who carefully simmers a sauce, the craftsman who restores old furniture, and the weaver who spins the wool for his tapestry. Craft skills require an intimate knowledge of the materials which are used in the realization of an objective. These skills atrophy in the expanding commodity environment and become a factor only in the production of luxury items, rather than a feature of everyday life for the majority of the population. Under the prevailing division of labour, through which the quantitative dimensions of the marketplace expand, most individuals could hope at best to develop craft knowledge in one or two domains, relying on the purchase of finished products for the remainder of their needs. With respect to the latter there is very little basis for an independent assessment of their qualities.

If attaining what I have called craft knowledge about the things that are supposed to satisfy our needs is impossible, what kind of knowledge can consumers expect to possess? They must be content to become acquainted with the claims advanced on behalf of commodities (by advertisers and other consumers) and to decide to accept - provisionally and arbitrarily - one set of claims rather than another. Given the sheer number of available items, the most unbiased package of information about commodities would be of little assistance, and consumer choice as a whole would remain a random activity.[6] Such 'choice' represents little more than subjecting one's body and mind to a grand experiment in the marketplace in order to determine whether any of the increasingly bizarre claims made on behalf of the products will prove to be valid. One-armed bandits in the dingiest of casinos offer better odds.

The consumer in this setting must contend not only with the number and variety of goods, but also with the intricacy of their composition. For only an advanced industrial order which applies the most sophisticated scientific and technological ingenuity to the tasks of mass production can deliver such a quantity of things; and both the materials and mode of fabrication for those things depend upon production processes so complex that few persons other than the holders of doctorates in physical chemistry have much hope of grasping them.

Recently an economist has offered a new justification for protracted schooling in the contention that 'education makes people into more efficient consumers.'[7] The holders of higher university degrees, armed with portable calculators and the latest *Consumer Reports Digest*, may have relatively better success than the high school dropout in negotiating a path through the jungle of commodities, but still their decisions will not be based on any significantly greater knowledge about the composition of the products themselves.

Craft knowledge about material things ought to be a decisive factor in the choices individuals make about the kinds of objects that are suited to their needs. We might call what is required competent personal knowledge or simply educated common sense. An individual's choices can fulfil his own desires only if the application of his native senses - the olfactory, visual, auditory, tactile, and gustatory capacities - and a modest understanding of the principles of organic chemistry, together with a reasonable amount of time for research and acquisition, will suffice for procuring material goods that can assist in realizing those desires. Where this is not the case, consumer choices are largely a process of becoming acquainted with messages about commodities and arbitrarily choosing which to believe for the time being. The accelerating rate of product turnover provides a clue to the shallowness of wants in the prevailing market setting: things that appear so indispensable one day, only to be discarded in favour of others on the next, cannot be presumed to stir the wellsprings of desire very deeply. However paradoxical it may seem, the fact is that one outstanding effect of the high-consumption lifestyle is to induce individuals to become more and more indifferent to each of their specific wants in proportion to the increasing amount of time and resources expended on the total activity of consumption.

There is another paradox as well: the less able one is to apply personal or craft knowledge in market choices, the more necessary such knowledge becomes if one is to avoid endangering one's health in the act of consumption. Continuous rapid progress in scientific research and technological application has enabled producers to flood the market with complex chemical compounds, including many toxic substances, the use of which requires the exercise of some care and attention on the consumer's part. All those destined for actual ingestion by humans have been first tested individually on animals, but there is as yet no adequate information on the potential long-range and combined effects on individuals and on the natural environment of the cycling of massive quantities of such substances through ecosystems.[8] Without them, however, the number and variety of commodities would be drastically reduced from present levels, not to mention future aspirations; therefore the present tendency represents a colossal auto-experiment by individuals, based on a naïve faith in the power of science to counteract any deleterious side-effects by means of further innovations and new products.

Some major problems already have become apparent - to list but a few: poor nutrition and a greater incidence of certain diseases (such as intestinal cancers) resulting from a high intake of processed foods; health problems associated with the use of cosmetics; a connection between food additives and hyperactivity in children; the dependence of a significant proportion of the population on regular dosages of prescription drugs, especially stimulants and depressants (as well as widespread dependence on alcohol); and the carcinogenic properties of many widely used chemicals.[9] Given the uncertain state of scientific knowledge itself with respect to these matters and the likelihood that this uncertainty will persist and perhaps even grow in the coming years, individuals will continue to experiment with physiological processes under conditions of inadequate personal knowledge. For example, a glance at the advertisements in any medical journal, which detail the potential side-effects associated with modern drugs, shows to what extent the average individual must put his faith in his doctor's expertise, while the doctor in turn is only relying upon the expertise of researchers in pharmaceutical laboratories. At every level - researcher, doctor, hospital, and patient - the chances of erroneous use multiply in proportion to the complexity and variety of the chemical ingredients; recent critical studies of medical practice show that the incidence of error is not insignificant.[10] Of course individuals are willing to take these chances in order to suppress particular ailments and prolong their lives so far as possible, but the potential aggregate effects of this behaviour with respect to human health and population genetics cannot be calibrated as yet.

The psychological risks inherent in a high-consumption lifestyle are no less important than the physiological ones. A study reported that the average consumer notices eighty advertisements per day and reacts (positively or negatively) to a dozen of these; but apparently the sheer quantity of messages produces a kind of behavioral dissonance, for other studies have shown no significant variations in actual purchasing patterns between those who distrust advertisements or maintain that they are not influenced by them, and those who find the messages credible and describe themselves as easily persuaded.[11] No single person can possibly perform all the experiments on his desires that a complex market demands; therefore manufacturers regard the consumer population as composed of 'pioneers' and 'sheep' (to use Robin Marris' terminology), with the former category consisting of those willing to venture into new product areas in the expectation that others will follow. Presumably each of us has some pioneering spirit, the aggregate of which keeps the entire flock moving.

What psychological risks are present under such conditions? I suggest that each aspect of a person's needs tends to be broken down into progressively smaller component parts, and therefore that it becomes increasingly difficult for the person to integrate the components into a coherent ensemble of needs and a coherent personality structure. For the sake of argument let us assume that everyone has a need for self-respect and respect from others. In responding to this need individuals cultivate certain internal dispositions or personality traits and also adjust their external appearance in accordance with changing social norms. Today the need for an acceptable external appearance is fragmented in association with the proliferation of products designed to satisfy it. Individuals treat their own bodies as objects made up of component units, each having its own demands. Hair, face, mouth, eyes, hands, axillae, neck, crotch, legs, and feet all require the application of specific and distinct chemical mixtures, which together will make one's body pleasing to others and therefore a means of winning favour. Each of these specific needs in turn can run through potentially infinite permutations in relation to the technological sleight-of-hand embodied in commodities: a deodorant undergoes metamorphosis from paste to powder to spray, is scented with synthetic resemblances to all the fruits of the earth, is packaged in a dazzling variety of shapes and sizes - and so on *ad infinitum*.

This fragmentation of needs is just another name for the fragmentation of personality: under increasing pressure from new external stimuli 'the centre cannot hold.' Or, more precisely, personal identity becomes a supple mould reshaped daily by the message mix. How significant a role the market syndrome plays in socialization patterns within this setting can be seen most clearly in cases of adult individuals who enter the system as emigrants from very different societies. A woman with whom I am acquainted came to North America from an 'underdeveloped' land as the spouse of someone who had emigrated from the same area many years previously. She spent the better part of her first year here alone at home watching daytime television – where program content as well as commercial breaks supply an endless stream of messages about commodities – in order to overcome her condition of bewilderment and social paralysis. Upon reflection one can appreciate how sensible was this method of acculturation. In effect, she was learning *how* to want.

Of course the socialization process in every society teaches individuals how to shape their desires and behaviour into the 'holistic' orientation that constitutes one's personality, into an integrated perspective for interpreting one's needs. What is unique to the high-intensity market setting is the necessity for being able to identify states of feeling systematically with appropriate types of commodities. The vast number and variety of material objects enjoins the person to break down states of feeling into progressively smaller components and instructs him in the delicate art of recombining the pieces fittingly. The 'wholeness,' the integration of the components, tends to become a property of the commodities themselves: one who drapes the latest sartorial splendour over a properly deodorized frame, and then treats his locks with an old-fashioned pomade rather than the newest lacquer spray, will quickly learn what it really means to be an inattentive consumer. The point of all this is quite simple: the fragmentation of needs requires on the individual's part a steadily more intensive effort to hold together his identity and personal integrity. In concrete terms this amounts to spending more and more time in consumption activities.

A few years ago Staffan Linder, a Swedish economist, wrote a learned and witty book about the problem of time in relation to consumption. He explored the implications of the elementary facts – both obvious and generally ignored – that consuming goods takes time and that an individual's time is limited. The major implication is that there is an increasing scarcity of time in a high-consumption economy, which contradicts the familiar notion that, as industrialism and worker productivity grows, the amount of free time available to individuals enlarges correspondingly. An unbridled economy is an imperious horseman: production and consumption are the twin spurs with which it goads those called to its service, and the free time of individuals is among those many delicate things trampled underfoot in its headlong rush. Linder refers to some of the anthropological literature on the tempo of activity in primitive societies, and Sebastian de Grazia has done some elaborate calculations on time allocations in the contemporary era which Linder has relied upon in part to support his thesis.[12]

According to Linder there is an 'increased goods intensity' in the pursuit of enjoyment: 'The clearest examples of pleasure that are on the increase will be found among activities based on the use of things ... Consumption is being accelerated to increase the yield on time devoted to consumption.' Greater production efficiency means that more things (and a higher value of things) are available for the same expenditure of real income; consumption of things must therefore rise to ensure that an equal benefit will accrue for each unit of time devoted to that pursuit. To give a crude example – assuming that one can consume $50 worth of things in each hour devoted to that activity (averaging the value of all commodities), $50/hour will be the 'yield' on consumption time. If productivity doubled, the yield would be $100/hour, and thus one would have to double the value of one's consumption in order to maintain the same yield on consumption time. So long as this does occur, there will be increasing pressure on an individual's time, for it requires time to use the objects purchased (television set, golf clubs, sailboat, camper, and so forth). Therefore in the search for enjoyment the tendency will be for activities requiring the use of material objects to crowd out those not involving them: 'Not all cultural pursuits will be hit equally hard by the increasing competition for time. This is because different cultural pursuits differ with respect to their potential enhancement by an increased goods capacity. To admire attractive cloud formations or to reflect on the world at large requires only the individual's time and no goods at all ... [This] category of pursuit will suffer particularly from the competition for time. Goods will become increasingly cheap and time more expensive, but these pursuits offer no possibility of substituting time for goods. More goods will not increase the pleasure derived, but actually reduce it.'[13] This is Gresham's Law applied to consumption: wants for ever greater numbers of commodities tend to depreciate all types of desires that are not dependent upon the consumption of things.

The aspects of the individual's situation that have been discussed give rise to a situation in which individuals become confused about the nature and objectives of their wants. This cannot be explained in a simplistic fashion as resulting solely from the influence of advertising, although modern advertising certainly is one decisive factor in our confusion about our needs. The images employed in the construction of advertising messages often incorporate a set of ambiguities about wants and their objectives. To take only one example, the most common form of automobile advertisement displays the vehicle against the background of a beautiful natural environment - near mountain ranges, oceans, rural settings, and other relatively unoccupied locations. The force of the overt message depends upon the powerful appeal of those settings to the crowded urban dweller, and the covert communication suggests the usefulness of the vehicle in making those settings accessible to the individual. If this were the primary reason for the purchase of an automobile for millions of consumers, of course, those settings would not remain uncrowded for long. In fact this is not what actually happens. One study showed that 93% of all automobile trips during the mid-1960s in the United States were made entirely or partially within cities.[14] The point is that in this process individuals become confused about their wants in relation to the means that are supposed to satisfy them.

On a more general level, popular activities such as tourism and camping reveal a deep-rooted state of confusion about means and ends in relation to consumption. The most common type of the former is the package tour, which has been designed expressly to cope with time constraints and the dictates of quantitative efficiency. Its features are familiar: the tight scheduling so as to encompass the maximum number of 'sights'; the special hotels where the language, service, and cuisine of the exotic locale has been adjusted to conform with the native tastes of the visitors; the organized dash through the storied palaces of high culture; the mass-produced souvenirs and projector slides which furnish evidence of a journey successfully accomplished (and which are often imported specifically for the tourist trade, as with the Canadian 'Mountie' figurines manufactured in Japan); and the banding together of the group, who deal with the local population only through professional intermediaries. Anyone strolling in a European city during an early summer morning has seen the collection of faces staring out of the innumerable tour buses and knows how long and arduous is these vacationers' working day. For the purposes of the package tour it is the individual tourist himself who has been 'packaged,' artfully wrapped and insulated against the possibility of ever making contact with a different culture; in a real sense he never leaves home, no matter how extensive the physical journey may have been.

The effort to make contact with 'unspoiled nature' by an excursion through the campgrounds in public parks has similar characteristics. The traveller insulates himself against the possibility of a qualitatively different experience simply by transporting with him as much of his regular complement of artifacts as he possibly can; for the most fortunate ones there is the enormous, self-propelled recreational vehicle equipped with the complete range of normal household appliances. The trained personnel staffing the campgrounds organize nature in much the same way as the tour proprietor organizes a foreign culture. There are short hiking trails through 'managed' forests, check-out times for the campsites, and evening shows in the amphitheatre where the range of potential experience can be efficiently compressed and interpreted for those who have no time to encounter it directly. Time is scarce, and there are so many sights to be glimpsed through the windows of fast-moving vehicles that traverse the highways linking one campground with another; for viewing those which as yet cannot be approached by mechanical conveyances a full array of projector slides may be purchased.

Activities such as tourism and camping betray a profound confusion within the wants and objectives of individuals. Experience is filtered through a many-layered veil of commodities - the physical accoutrements of travel, mass-produced mementoes, planned menus, and interpretive portfolios indicating what should be seen - which refine the richness and diversity of both the natural environment and human culture into standardized states of feeling. In effect the material objects simulate experience.[15] Filtered through layers of commodities, the natural and human environment is progressively simplified, more smoothly ordered; the abrasive particles which might disrupt the flow of everyday normal experience - which might stir modes of feeling not dependent upon the acquisition of things - are trapped and removed.

This constricting of experience is a clue to the possible future development of all industrialized market societies. For what is so striking about this social practice wherein needs and commodities are elaborately linked together is that the material extravagance now associated with it in the industrialized nations ultimately may prove to be unnecessary. The simulation of experience that at present occurs through the medium of commodities can be looked upon as a training exercise or a Pavlovian conditioning process. The material object in which the message is embodied may be dispensed with after a suitable interval, when the message alone will be capable of ministering to the want. Martin Krieger has presented the structural dynamics of this process with great clarity. His discussion of the way in which society can meet the demand for natural environments provides an insight into more general possibilities: 'The supply of natural environments is affected by technology in that it can manipulate both biological processes and information and significance. The advertising that created rare environments can also create plentiful substitutes. The supply of special environments can be increased dramatically by highlighting (in ways not uncommon to those of differentiating among groups of equivalent toothpastes) significant and rare parts of what are commonly thought to be uninteresting environments ... It is likely that we shall want to apply our technology to the creation of artificial environments. It may be possible to create environments that are evocative of other environments in other times and places. It is possible that, by manipulating memory through the rewriting of history, environments will come to have new meaning. Finally we may want to create proxy environments by means of substitution and simulation. In order to create substitutes, we must endow new objects with significance by means of advertising and by social practice. Sophistication about differentiation will become very important for appreciating the substitute environments. We may simulate the environment by means of photographs, recordings, models, and perhaps even manipulations in the brain. What we experience in natural environments may actually be more controllable than we imagine. Artificial prairies and wilderness have been created, and there is no reason to believe that these artificial environments need be unsatisfactory for those who experience them.'[16] A recent radio news report told of a multi-storey parking facility for camper vehicles in a Texas city, where each vehicle stall is carpeted with plastic grass, and motion-picture images of wilderness scenery and wildlife are projected onto the surrounding concrete walls. The report may be apocryphal, in which case it may only have anticipated the event.

This perspective is encouraging in its own perverse way. For in the end we may not have to worry overmuch about potential shortages of materials and energy. The high-intensity market setting would serve as a mechanism for the education of desire. With the population trained to simulate their experiences through the many-layered veil of commodities, complex images of things gradually could be substituted for the present elaborate network of things and messages. After a sufficient amount of training in this social process, individuals may find that a stimulus such as that provided by the notorious 'feelies' in Huxley's *Brave New World* duplicates satisfactorily an entire world of experience.

THE INSATIABILITY OF WANTS

In modern social theory one often encounters the proposition that human wants or needs are insatiable. (Later in this essay I will take up the question whether it is wise to distinguish between 'wants' and 'needs.') The proposition is repeated with great frequency in writings on economics and psychology. This example is from an economics textbook: 'For all practical purposes, human wants may be regarded as limitless. An occasional individual may have everything he wants, but man's capacity to generate new wants as fast as he satisfies old ones is well-known psychologically.'[17] The leading contemporary authority in the field of consumer research in the United States puts the point this way: 'The American, confident of his own power to advance his well-being, steps up his wants at an equal pace with, or even faster than, his accomplishments.'[18]

If we simply accept this proposition as a reasonable statement of our situation, instead of examining its implications carefully, we appear to be faced with an insoluble dilemma. For no matter how wealthy and productive our society might become, we would always require higher levels of production and greater quantities of goods. The projections for material demands that were referred to in the preface give some idea of the staggering requirements that we will face if we proceed along this path. Under these conditions all of our social concerns will have to be concentrated on the single objective of organizing a productive capacity that can keep step with the escalation of our demands.

In most discussions of this issue it is taken for granted that the term *insatiability of wants* refers to wants for commodities – that is, wants for material things and the associated services that are necessary in order to use and enjoy them. The commodity is understood as a means for the satisfaction of a desire, serving as a link to unite an impulse with its objective. However, to specify the nature of this link and how it functions in the everyday activity of the consumer is a very difficult undertaking. The major fault with most characterizations of the supposed insatiability of wants is that they do not attempt to clarify this linkage at all. In all social settings wants and needs represent complex states of feeling for individuals, and in an advanced stage of a market economy the linkages between desires and types of satisfactions are highly ambiguous.

Most contemporary discussions fail to explore the problems that individuals face in interpreting their needs when the market economy expands to the extent which now prevails in the richer industrialized nations. When individuals switch their pattern of needing from the context of a relatively modest and slowly changing assortment of goods to one which makes available an immense and rapidly changing variety, they must reinterpret the entire pattern. For although food items, for example, would be present in both contexts, the qualities of such items are very different, and each item stands in a different relation to the entire assortment. The degree of familiarity with each item - what I have called the craft knowledge concerning the qualities of things - necessarily diminishes, along with the capacity to make an independent assessment of its suitability for one's needs. On account of this there is the danger that individuals will misinterpret the relationships between their needs and the available means of satisfaction for them.

An example of the way in which this occurs is provided by the problem of obesity, which has been studied for some time by social psychologists. Obesity resulting from an excessive consumption of food has been related to states of anxiety and depression, and these in turn are thought to be rooted in the individual's failure to achieve satisfaction of the needs for acceptance, approval, and achievement.[19] The deflection and distortion of these social interaction patterns into overconsumption of food means that obese persons become relatively unresponsive to internal physiological determinants of the need for food and react instead primarily to external cues in the environment. In this instance we have an extreme case of persons who misinterpret their needs and damage themselves accordingly. The obese person remains fixated on one set of wants and interprets all of the messages, cues, and stimuli in his environment in terms of a single overriding mode of gratification, instead of following the normal pattern and distributing the objectives of his wants among different sets of commodities. A classic imaginative portrayal of this process is Mr Joyboy's mother in the film version of Evelyn Waugh's *The Loved One*, who indulges in actual gluttony while simultaneously deriving vicarious satisfaction from televised food commercials.

In a high-consumption society individuals must depend upon social cues to guide them in deciding how extensively to use any individual commodity or sets of commodities in relation to the others that are available. The man who douses himself liberally from head to foot with a single all-purpose lotion will be esteemed less highly than someone who selects the correct combination and amounts of hair-spray, after-shave lotion, mouthwash, talcum powder, cologne, and various deodorants. Moreover, he would depend on these cues first to know how to vary the assortment according to the occasion (for what is requisite for a formal dinner is inappropriate for a fishing expedition) and second to know when changing social tastes dictated enlarging the assortment or replacing it entirely with 'improved' products.

The pathological state of consumer behaviour, for example the over-consumption of food, results from refracting all interpersonal and market stimuli through a particular fixation. The normal state is the ability to reshuffle the distribution of expected satisfactions regularly in accordance with changes in the quantity, variety, and relative importance of commodities. Since satisfaction is a private matter and so cannot be measured, there must be some public standard of normalcy and rationality for market behaviour; the switching of wants within certain parameters as new commodities appear supplies the standard. We recognize an abnormal condition in the case of obesity because the overtly quantitative aspect of the problem is so obvious; but so long as the individual remains just moderately overweight, for example, in the process of switching his desires for food according to each successive innovation in packaging, processing, and so forth, we can say that he is exhibiting normal behaviour. The striving for satisfaction, which is an 'intensive' dimension of experience involving the internal disposition of the person, has an 'extensive' dimension with respect to the willingness of individuals to adjust their needs to an enlarged assortment of goods.

The proposition about the insatiability of human wants is ambiguous primarily because the relation between needs and commodities in our society is so ambiguous. It is difficult to know what is meant by the term *satisfaction of wants* in this setting. Here the complexity of the commodities that are produced in an advanced industrial society is very significant. They are not simple things, but rather an unstable unity composed of many characteristics, qualities, and 'messages' (this point will be discussed in more detail later). The problems which arise in this setting can be grasped if one compares the modern consumer's situation with that of a shopper in a more limited market economy. In eighteenth-century newspapers, for example, one can read merchants' advertisements which announce that a ship has arrived and that various goods are available - tar, nails, tea, cloth, and so forth. The qualities of such things are quite apparent and could be assessed at that time by the average individual's craft knowledge; in other words, the link between needs and commodities in this setting is relatively simple.

In the high-intensity market setting, on the other hand, both the states of feeling that are incorporated in an individual's wants and the multidimensional aspects of commodities are highly complex; the complexity of the interplay between needs and commodities increases exponentially as a result. It is far too simplistic to adopt the conventional description of this process as one in which 'new' wants emerge attendant upon the 'satisfaction' of previously existing ones. In this setting wants become less and less coherent, and their objectives less clear and readily identifiable, as individuals continually reinterpret their needs in relation to the expanding market economy: 'The modern adaptive consumer not only participates intensely in market processes but also changes his behavior according to the requirements of the situation, and thus facilitates and even anticipates new technologies and styles of life.'[20] With the fragmentation of needs into progressively smaller components and the regular reshuffling of these components into momentary desires or states of feeling that are discharged in consumption activity, the determination of the extent to which the satisfaction of wants is being accomplished at any point becomes problematical.

New commodities, which today appear steadily and in great numbers, simultaneously promise the satisfaction of wants and promote a feeling of dissatisfaction with regard to the previously existing array. The dizzy pirouette of wants and commodities presents to the individual an ever-changing ensemble of satisfactions and dissatisfactions in terms of which there is no resolution, but only a continuous movement from a less extensive to a more extensive participation in market activity. In other words, the individual is encouraged to shift his ensemble of satisfactions and dissatisfactions continually from a smaller to a larger set of commodities (or sets of commodities with more elaborate characteristics).

The colossal productive system whose impetus is derived from this interplay of needs and commodities appears at first glance as a kind of self-inflating hermetic sphere: one must search diligently for a point of entry that would enable one to distinguish clearly the intended objectives of the wants that are comingled with such an incredible variety of material objects. Contemporary theories of needs and of consumer behaviour have not attempted to develop a critical perspective on this system. The theory of the interplay between needs and commodities presented in Part Two has this objective.

Society

What are the main institutional problems that must be faced by any society which includes the kind of market setting that has been described in the preceding section? Such a society encourages its citizens to orient their search for the satisfaction of their needs more and more exclusively toward consumption activities, in part by neglecting all other possibilities for individual self-fulfilment (such as participation in creative and satisfying work environments). It must therefore devote its energies primarily toward ensuring that adequate material means are available for its constantly expanding economy. When the dimensions of the market economy have grown to the point which now prevails in the richer industrialized nations, public concern over the possible scarcity of resources and energy becomes a familiar feature of everyday life.

The productive system which sustains a high-intensity market setting depends upon highly centralized organization and systematic planning in both public and private sectors. In all present-day capitalist societies there is a general consensus that government should assume an important share of responsibility for regulating the use and development of vital resources and the tempo of economic growth. There are many contentious issues raised by this activity; I shall confine my remarks to those which are directly relevant to the main theme of this essay.

THE THREAT OF SCARCITY

In a college textbook published a few years ago we find this account of our situation: 'Scarcity: a condition which mankind perpetually faces because at any moment in time total resources are fixed.'[21] In this formulation, which is often repeated with minor variations, scarcity appears to be an inherent feature of the human condition as such. It locates the source of our dilemma in the relation between human beings and available resources, and it implies either that nature has failed to endow our planet richly enough or that we cannot gain access to sufficient resources with the means currently at our disposal. In either case this is a one-sided and somewhat arbitrary formulation of the matter. For scarcity may be regarded from another viewpoint as stemming from an allocation of abundant resources in a wasteful, unjust, and irrational manner. In this view scarcity is a socially created condition that arises out of a particular organization of productive activity. The threat of scarcity is a permanent feature of our present society, no matter how vast the supply of available goods, because the escalation of material demands only plunges individuals more deeply into the ambiguous ensemble of satisfactions and dissatisfactions – and thus into greater confusion about the relationship of their needs to the many dimensions in the sources of possible satisfaction.

Some of the most influential proponents of the expanding market economy in the early modern period laid the foundations of the widespread belief that scarcity is rooted in the limited productivity of the prevailing economy. They argued the superiority of complex market economies and painted a dark picture of earlier systems. Hunting-and-gathering societies, for example, were portrayed as being perpetually on the verge of general starvation, with the individual in those societies devoting every waking hour to an exhausting search for the barest minimum of subsistence. According to this view humanity barely succeeded in escaping this hour of its peril into the settled agricultural state, where exchange and the market flourished, and where the painful but uplifting march of civilization began.

At the very least this picture was overdrawn – and it may be a complete misrepresentation of the real situation. The anthropologist Marshall Sahlins, in a provocative essay entitled 'The Original Affluent Society,' presents a case for reversing the interpretation, assembling evidence that the hunter-gatherer societies maintained a quite adequate level of satisfaction under normal circumstances and did so during a working-day less extensive than our own. This is not the place to evaluate the contending interpretations; owing to the paucity of evidence, there will probably be no definitive conclusion to the debate. But Sahlins' essay has the more general objective of reopening the whole question of scarcity in human society, and he tries to turn the tables on the champions of the modern way: 'The market-industrial system institutes scarcity, in a manner completely unparalleled and to a degree nowhere else approximated.'[22]

What does this mean? Sahlins argues that scarcity is 'a relation between means and ends'; in effect, *felt* or *perceived* scarcity is the only reliable guide to assessing the degree of success or failure involved in the social arrangements whereby human groups organize the attempted satisfaction of needs. If we view scarcity as the disparity between our wants and our capacities, we can understand the possibility that scarcity might increase simultaneously with rising social wealth and productivity.

In this perspective scarcity is not a simple relation between resources and wants, but a far more complex correlation between the interpretation of human needs through particular forms of social interaction, on the one hand, and, on the other, the extent to which the material aspects of the surrounding environment become the conscious object of the attempts to satisfy those needs. Accordingly we would not assume that people always articulate their needs in the same terms, nor that they always regard the natural environment primarily as the repository of a greater or lesser quantity of resources. This assumption is quite different from the one to which we have become accustomed - namely, that the human condition is characterized by a perpetual scarcity of resources relative to wants.

How can this alternative assumption be applied to the situation of the richer industrialized nations, with their growing fears of future shortages and scarcities? How does this society create the experience of scarcity in everyday life? This is a complex and subtle process. For example, let us assume that our society requires a high degree of mutual respect among persons in order that the innumerable daily interchanges upon which it depends can occur smoothly; correspondingly, we could postulate a need for interpersonal respect in all individuals living in such a society. How can this social and individual need be satisfied? At the risk of oversimplifying matters I shall consider only two approaches and depict each in extreme form in order to clarify the point at issue. At one extreme, we can imagine a society that had evolved socialization processes in which individuals developed personal self-identity and mutual respect through introspective meditation, play rituals not confined to the period of childhood, educational experiences in structured but non-hierarchical settings, and opportunities for creative self-expression in work environments. At the other extreme, the social order might closely identify the search for personal self-identity and interpersonal respect with the purchase of commodities - clothing, grooming aids, deodorants, private automobiles, university degrees, personality improvement techniques, and so forth, all renewed regularly (and at an accelerating rate) under the régime of fashion. The basic point is that under both sets of circumstances interpersonal respect might remain scarce to some extent, but in the second case this scarcity will be compounded by scarcities of the many other just-mentioned things which have been associated with it through the commodity-oriented socialization process.

There is a 'psychology of scarcity' associated with the multiplication of material goods as instruments for the satisfaction of needs.[23] This arises from the kinds of social interaction whereby individuals encourage each other to believe that failure to procure the means for purchasing certain goods will exclude them from all hope of satisfying the needs which have come to be associated with those goods. And to the extent that the majority of individuals participate in this process, the sources of satisfaction for any need will be as scarce as the particular commodities in question.

Experienced scarcity may arise along the entire continuum of possible supply. Even the most elaborate provision of material means directed toward a particular need offers little assurance that the need will be answered effectively. The facilities for personal mobility in North America are a case in point. At present there are half as many private automobiles as there are individual persons; of course these are unevenly distributed, but even if it were possible for every individual to own a private vehicle, this would not in itself assure any of them of the requisite mobility. The reason is that the demand for mobility is determined by the social arrangements as a whole, for example the physical separation of work-place and dwelling-place. The overall arrangements may make personal mobility progressively more difficult (i.e., more scarce) in proportion to, or even more rapidly than, the expanding aggregate supply of private and public power-driven vehicles.[24] The complexity of the overall structure deprives individuals of the ability to rely upon directly accessible means (such as their own physical capacity for mobility) for the effective discharge of their required functions.

The hidden dilemma for any society which attempts to sustain a high-intensity market setting is, simply stated, that even a steadily increasing array of commodities cannot alleviate the threat of scarcity as it affects the everyday lives of individuals. The dilemma is rooted in the fundamental characteristic of this setting, namely the tendency to orient all needs toward the realm of commodity consumption.

Commodities are by definition things that are relatively scarce. Thus when the sphere of the satisfaction of needs becomes more and more closely identified with the sphere of available commodities, the potential range of the experience of scarcity for individuals will increase in proportion to the number of commodities. At every moment the assortment of goods is, then, simultaneously the assortment of socially induced sensations of scarcity. This kind of society is beset with a permanent contradiction between expanding wealth on the social level and intensified experiences of scarcity on the individual level. No matter how assiduously (or how successfully) it mounts the search for increased supplies of resources and energy, it cannot resolve the contradiction.

THE MANAGEMENT OF DISCOMMODITIES

In the processes of production and consumption matter and energy are cycled through the physical environment. The resulting amount of 'residuals' (gases, particulates, thermal pollution, solid wastes, and so forth) has reached immense proportions, especially of course in the industrialized nations.[25] Existing institutional frameworks, such as property laws, are slow to adjust to the new problems in this area. A recent economic study concluded that 'we persist in referring to the "final consumption" of goods as though material objects such as fuels, materials, and finished goods somehow disappear into a void - a practice which was comparatively harmless only as long as air and water were almost literally "free goods." '[26]

We have always depended upon the surrounding air and water to dissipate and recycle wastes. There is thus a dual human dependence on the natural environment. Increasing material demands alter our perceptions of our relationship to the environment, so that we tend to regard the rest of nature almost exclusively as a warehouse of resources and a dumping-ground for wastes. This means that the threat of scarcity has two dimensions, for at present rates of increase 'humanity will very soon be faced with two sorts of irremediable scarcity, those arising from the finiteness of natural resources and from the assimilative capacity of the biosphere.'[27] The second dimension of potential scarcity poses a particularly complex set of problems for an industrialized market society.

Commodities are the material goods and associated services that are the desired objects of individual wants; the acquisition and use of them supposedly is the means for the satisfaction of needs. As noted above, the production and consumption of commodities necessarily produces discommodities, which are the residuals or waste products that must be dissipated in the environment. Industrial production, utilizing vast quantities of new substances derived from scientific and technological research, supplies the range of new commodities that maintains the impetus behind the high-consumption ideal. Industrial production also results in vast quantities of residuals: gases (CO_2, SO_2, etc.), airborne particulates, hydrocarbons, phosphates, insecticides and herbicides, contaminated water, and many others. When they are generated in large quantities they do not all dissipate freely and safely in the environment; they become discommodities, that is, undesired things which give rise to costs when we dispose of them. These costs may increase the prices of commodities directly, when pollution control equipment is installed in factories, for example, or they may increase public expenses such as water treatment at tax-supported facilities.

Society must therefore accept responsibility for the management of discommodities through a variety of institutional responses, in order to avoid the otherwise serious health problems that would arise were these dangerous residuals simply to be dumped into the environment. A society which wishes to sustain a high-intensity market setting must manage discommodities in such a way as to maintain a high level of utilization of commodities by the population as a whole. This process has many aspects; I will comment on only two of them for purposes of illustration - the separation of commodities and discommodities and the tendency to confront only the short-term dimensions of the situation.

Some of the problems associated with the individual's knowledge about commodities have already been discussed. A related aspect of this matter involves the individual's knowledge about discommodities and about the relationship between commodities and discommodities. In the first place, the complexity of modern industrial processes and the rapidity with which new chemical compounds are put into use means that considerable time may elapse before any scientific knowledge is possible about potential hazards resulting therefrom. This was the case with the recent discovery of dangers associated with the vinyl chloride compounds used in making plastics. The persons most immediately affected in these circumstances are the production workers who are exposed daily to toxic substances. The separation between the two spheres here involves the difference between higher risks to those engaged in production and the general benefits of increased consumption for the entire population.

So far public controversy has revolved around the carcinogenic and other effects of substances such as asbestos, uranium, and vinyl chlorides on the production workers in various industries. There may be general environmental hazards affecting the entire population, however, that are as yet unknown or only suspected. One such case which received widespread attention recently was the possible effect on the atmospheric ozone of the gases used as propellants in aerosol containers. The separation between the immediate level of benefit for the consumer in the use of aerosol sprays and the potential environmental dangers is compounded by the inherent difficulties in obtaining adequate scientific knowledge about the situation. A desire not to interrupt the production processes usually leads to an insistence on demonstrating 'conclusive' scientific evidence of a hazard, but such problems are so vast and complex that the evidence may be forthcoming only after it is too late to devise a remedy or only after the costs of alleviating the situation have increased dramatically.

The economic structure of multinational corporations makes possible another kind of separation between commodities and discommodities. In response to more rigorous restrictions on the emission of pollutants in the advanced industrial societies, the corporations have begun to transfer some production operations to less-developed nations which have no such restrictions. This has occurred in asbestos manufacturing, for example.[28] The finished products are consumed almost entirely in the industrialized societies, and the population in the country of manufacture suffers the environmental hazards in return for whatever economic benefits are derived from the transfer of the processing operation. The cost of manufacture may even be reduced if cost savings due to differences in prevailing wage rates are greater than the additional shipping or other charges. At any rate those who enjoy the goods need not experience (at least for the moment) the environmental effects of the production residuals.

The discommodities resulting from modern industrial processes are often complex substances; reliable information on their environmental impact is non-existent in most cases and is in any event difficult to obtain. Air and water pollution resulting from human activities is not new. What is new is the complexity and variety of recently developed chemical compounds and the rapidity with which they are introduced into large-scale production processes. The more obvious short-term pollution problems can be contained within acceptable parameters by legislation and public concern. But the management of discommodities usually involves increased costs of some kind, costs which eventually will be expressed in higher prices for commodities. To maintain our present course the long-range, currently unknown environmental hazards resulting from modern industrial production will tend to be underplayed in order to minimize short-term costs.

In one of the preceding sections of this essay it was suggested that individuals in this market setting experiment with the psychological and physiological aspects of their needs in responding to the regular appearance of new commodities. The ways in which contemporary society manages discommodities show that these individual experiments are matched on the social level by a general experimentation with respect to potential environmental hazards. According to the prevailing wisdom we must have demonstrated proof that a danger exists before we restrict the use of an offending substance in our production and consumption activity. The canons of proof that obtain in the scientist's laboratory, however, may be inappropriate for our practical assessment of a substance's impact on the environment. Neither society nor the biosphere should be looked upon as a laboratory in which we can experiment freely with our desires while paying no attention to the possible harm that might accrue.

Now that some degree of public awareness concerning these hazards has been developed, we can expect that the attempt to maintain the separation between commodities and discommodities will become a progressively more complex social problem. The recent economic recession shows how fragile are the environmental standards that are supposed to define acceptable pollution levels. At the first sign of economic trouble the cry is raised for the relaxation of those minimal safeguards which have been in effect (whether enforced or not is another matter) for a very short time. The political and economic interests that maintain the high-intensity market setting exert powerful short-range pressures against whatever appears to threaten the viability of that setting. Estimates of the long-range environmental impact of residuals and the economic cost of discommodities have to be based on woefully inadequate information; at present the countervailing pressures based upon concern about environmental hazards are very weak. One cannot expect that in this setting individuals will be encouraged to understand the relationship between commodities and discommodities and to reassess their wants in the light of an understanding of that relationship - in short, to consider their wants in the context of the potential environmental risks that arise in the production and consumption of present-day commodities.

For the foreseeable future the management of discommodities will be governed by a set of environmental standards that are formulated in the interplay between scientific knowledge and short-term political expediency. This follows from the fact that steady growth in the provisioning of commodities remains our overriding social objective. Whether and in what measure more adequate knowledge about long-range environmental factors will occasion a rethinking of that objective is impossible to determine now.

Non-human nature

The problem of discommodities and residuals in modern industrial societies provides the clearest indication of how important it is for us to regulate our relationship to the natural environment with some sensitivity to its requirements, as well as to our own. The greater are our demands upon the environment's resources, the greater is our dependence upon its capacity to adjust to those demands. Instead of attempting to understand the character of our increased dependence on the regenerative capacity of the biosphere, however, we have grown accustomed to proclaiming our independence of the kinds of limitations on human activity that were operative in pre-modern times. The belief that man has achieved mastery over nature through modern science and technology has encouraged a spirit of indifference toward the common interests of many living things on the planet for which we may have to pay dearly in the future.

This spirit of indifference toward the ecological context of human activity pervades almost all modern discussions of human needs. The structure of human needs has been explored intensively with reference to the individual's psychological dynamic and the social bonds among individuals, but not with reference to the relationship between man and the rest of nature. Only in anthropological studies is the ecological context often made apparent.[29] But it is wrong to believe that this context is relevant only to the practices of primitive human societies. One cannot fully understand the character of human needs in any social context, including the high-intensity market settings in contemporary industrialized societies, if one ignores its third dimension -

the relationship to the world of non-human nature that is implicit in every human social organization. If we look carefully at some of the attitudes toward nature which have been influential in modern history, we can achieve a better perspective on the problem of human needs in our society.

THE CONQUEST OF NATURE

One of the most powerful currents of thought since the seventeenth century has been the conviction that the human species should and will 'conquer' nature through the progress of modern science and technology.[30] Its imprint can be found in the daily newspaper and the politician's rhetoric as well as in philosophical tomes. Its vitality flows from a solid bond between theory and practice. The conquest of nature consists not merely in the elegant formulas of mathematical physics, but also in a demonstrated practical mastery over the natural environment. Science, technology, and industry working in tandem, and not any one of these by itself, realize the human conquest of nature – necessarily so, for according to the most influential interpretation of this enterprise, its objective is eminently practical: the satisfaction of human wants on a hitherto undreamed-of scale. Or, in Francis Bacon's magnificent phrase, the 'relief of the inconveniences of man's estate.'

The secular idea of domination over nature was borrowed from an older conception. As a representation of the relationship between man and the rest of nature, it originates in one tradition within Western religion and philosophy, where spirit is thought to rule matter absolutely and where mankind is regarded as having exclusive rights of vassalage with respect to the living things of the earth.[31] This religious and philosophical perspective was based upon a conception of a mode of reality separate from nature (the realm of spirit) in which mankind participates. Non-human nature merely supplies the stage props for a spiritual drama whose cast is exclusively human. When this perspective was secularized, when the domination of nature by man came to be regarded as the outcome of his own scientific and technological ingenuity rather than as a gift from God, a radical flaw in its logic was concealed. This flaw is the presumption that the secular process of the control of nature by man – an accomplishment arising from the productive rationality of science, technology, and industry – was itself under rational control. That presumption is indispensable for the notion of domination over nature. For would it not be absurd to think that an agent which was not the master of its own nature could exercise mastery over its environment? Bacon, who saw this difficulty and recognized its seriousness, seized upon the following expedient: 'Only let the human race recover that right over nature which belongs to it by divine bequest, and let power be given it; the exercise thereof will be governed by sound reason and true religion.'[32]

But what if religion were to prove unequal to the task? What then would be the situation of those human creatures who in Bacon's eyes are 'full of savage and unreclaimed desires' and who would now command the powers of nature in ever greater measures?[33] Through the secular mastery of nature the material conditions of human life would be changed dramatically, but the propensities of human nature itself would continue to run their established course. The unstable balance of forces governing the relationship between this species and its environment would shift suddenly with unforeseeable consequences, for of course the *social* process through which human technological capabilities swiftly increased was not rationally directed in any way.

According to Bacon's influential argument, the 'inconveniences of man's estate' persist owing to humanity's traditional unwillingness to recognize that it possesses the capacity to enslave nature in the service of human ambition.[34] Bacon did not assert that the failure to achieve mastery over nature was the cause of human misery, nor did he claim to know what was the cause. He did contend - and in this he has won legions of supporters - that, whatever might be the sources of those inconveniences, mastery over nature could provide the remedy for them.

One of the main purposes of winning mastery over nature, which in practice amounts to ensuring the unimpeded growth of technology and industry, is to maximize the supply of goods available for the satisfaction of wants. This conception does not, therefore, merely extol the power of organized human ingenuity to shape its environment, but also affirms a particular outlook on the character of human desires. The latter was articulated during the same period by another tradition of modern social theory, inaugurated by Hobbes, wherein insatiable desire and perpetual scarcity are said to be the hallmarks of human social behaviour.[35] Insatiable desire, rooted in the social nature of the human species, means that the conquest of nature has no determinate objective, no inherent terminus. This conception suggests that we are embarked upon an endlessly enlarging spiral wherein increased control over nature strives to keep pace with burgeoning desire. There is a fundamental paradox here, for this conception describes a species that supposedly cannot exercise any rational control over its own behaviour, while simultaneously announcing its intention to subject all of non-human nature to its unchallengable control through its very rational science and technology.

If we accept this picture of human nature, we will be tempted to explain the preceding stages of social development as a series of expedients designed to cope with the inadequacy of available resources relative to wants, and to see the superiority of present-day arrangements in the industrialized nations in terms of their explicit dedication to maximizing the supply of goods. In other words, we would understand 'progress' in this sense as our ability to follow whither they take us the endless material demands arising out of our social nature. Therefore in a manner of speaking we would be only 'submitting' to our nature. A peculiar act, is it not, for the species which strives so mightily to dominate nature? Are we to believe that the gigantic enterprise upon which modern society has embarked resolves itself into a conundrum? That the reason for pursuing human control over all the rest of nature is to enable us finally to unbridle our own nature, so that it alone of all existing things is 'free'?

At first glance these questions may appear to represent nothing more than a play upon the ambiguities of the word *nature*. But I think there is more to the matter than this. Taken together, the conceptions of mastery over nature and insatiable desire are the abstract expressions of a concrete, developing social practice. They express a particular orientation of human activity in its relationship to the natural environment which, in more explicit terms, encourages individuals to regard nature exclusively as a support system for human wants. This perspective establishes the provisioning of our material demands as the single organizing principle for our relationship to the rest of nature. It may not seem to be very significant. However, I do think that by isolating it in this way we can see more clearly the novelty of modern society. For the first time there arises a social system which (in its contemporary form) methodically ransacks the earth in search of adequate resources and even casts an envious glance at the untapped bounty that may possibly exist on other planets.

NATURE: MEANS AND ENDS

Marx claimed that under the impact of modern capitalism 'for the first time, nature becomes purely an object for humankind, purely a matter of utility; ceases to be recognized as a power for itself ... '[36] He argued that this development was one of the 'civilizing' influences of capitalism, in that all the older forms of nature-worship, which played a part in the functioning of repressive social institutions, had been discarded. In other words, traditional institutions (such as the political order) had been defended as integral parts of the 'natural' order of the cosmos, as elements of a total human situation which could not be altered.

As it developed over a period of centuries capitalism dispensed with all kinds of justification for a political order based upon a vision of natural order.[37] Political institutions were no longer defended on the grounds that their structure was part of nature's. Gradually nature came to be viewed as purely a system of matter in motion governed by universally applicable principles. In this view there was no inherent purpose within nature; therefore there was no possibility that man could violate the natural order by turning it to his own purposes through scientific and technological discovery. The extrusion of all except human purpose from the cosmos meant that for the first time nature could be approached by man from a purely utilitarian perspective.

By *utilitarian perspective* I mean some kind of conscious organization of activity undertaken for the satisfaction of needs. The fashioning of tools is the most general form of such activity. The utilitarian perspective has always been a factor in human evolution, for tools are found among the early remains of hominid species uncovered in archaeological diggings. Every pre-modern society, however, developed symbolic categories that effectively put limitations on the appropriation of nature for human use, the best illustration of which is the 'sacred' places reserved for the gods. The utilitarian perspective was inhibited or limited to varying degrees by this means. All such limitations were removed when the emerging image of nature as devoid of purpose was joined with the social practice of capitalism, which repealed existing limitations on economic activity (such as the exclusion of labour and land from most exchange relationships) in accordance with a developing vision of society's goal as the maximizing of material productivity through a self-regulating market economy. The tendency of this development was to orient human understanding and activity in such a way that the natural environment could be treated as nothing but a repository of resources potentially available for use.

Once again Francis Bacon's writings offer a remarkably clear illustration of how the intellectual aspect of this transformation was accomplished. Bacon distinguished three kinds of ambition: personal, national, and species ambition. The first represents the competition for power, rank, and wealth carried out among individuals within a particular society; the second is the struggle for priority in the international company of nation-states, pursued through trade and warfare; and the third is the universal human quest for mastery over nature. The three drives, he contended, could be ranked in order of their suitability as human endeavours. Thus personal ambition is essentially vulgar and unworthy of encouragement. National ambition he approved of only with qualifications, for in a sense it is little else than the personal sort writ large. But the third, he intimated, is 'pure,' the only form of ambition that we can regard with unqualified approbation from an ethical standpoint.[38]

Why? On what grounds should the species ambition represented in the human conquest of nature be accorded this exalted status? Bacon does not reply to this question directly, but an answer is implied in his remarks. Under normal circumstances whatever successes are achieved as a result of either personal or national ambition are attained to the detriment of other persons or states. The pursuit of self-interest and national interest is in Bacon's view a zero-sum game; because of the harm done to those at whose expense the victors have triumphed, the results are morally tainted by feelings of guilt and shame. Opposing interests in domestic and international politics have the same ethical 'right,' and thus it is impossible to plead the cause of one against another. But should there be a sphere of activity in which only one interest is represented in the field of play, ambition would be free to follow its fancy unencumbered by the torments of conscience. That sphere does exist, according to Bacon, in the activity known as the conquest of nature.

Non-human nature has no inherent purpose and thus no interest, according to this conception, and in attempting to bend its ways to human purposes we have a morally legitimate enterprise for our species ambition. The aggressiveness involved in human ambition, which must be repressed in social contexts, can be turned loose against the environment and find release without the sense of guilt. Much later William James gave an explicit statement of this Baconian scenario. In his famous essay, 'The Moral Equivalent of War,' he recommended that the aggressive drives which traditionally had found an outlet in warfare be sublimated by requiring a stint of arduous manual labour in socially useful enterprises for all young men, who thereby would have 'done their own part in the immemorial human warfare against nature.'[39] Non-human nature 'pays the price' for achieving peace and serenity in human society.

What price? The whole point of Bacon's argument is that the conquest of nature is the one mode of expression for human ambition which entails no 'cost.' Nature is the malleable substrate in which human ingenuity works its wonders, casting matter and energy into moulds for desired objects. Bacon argued in effect that the intensive human transformation of the environment could absorb and neutralize those destructive passions which men otherwise turn against each other.

Is this indeed possible? Is it reasonable to suppose that the character of the specific drives which motivate the reckless, uncontrolled remaking of our earthly habitat is a matter of indifference? That the sphere of non-human nature, although enjoying no inherent purposiveness, contains an elixir capable of converting base human purposes into the material foundations of social tranquillity? My intention in posing these questions is not to ridicule the idiosyncrasies of an acute and penetrating seventeenth-century thinker. Bacon explicated, more clearly than anyone else before or after his time, the underlying logic of a social practice that has become known as the technological fix, or the conversion of political and ideological problems into technical ones. But every fix is a temporary palliative, as the addict knows, leading to an ever-greater tolerance and dependence; and a larger dosage is required to produce the same effect over time.

If (as Bacon thought) human ambition is charged with destructive energy, which when released within society breeds bitterness and disharmony, can its negative consequences be overcome by deflecting that energy into the conquest of nature? This is most improbable in my view. It would be more reasonable to expect that the character of the drive will be embodied in the result; that inevitably the avarice, cruelty, and ugliness inherent in this ambition will not disappear through absorption but rather will be reflected in the physiognomy of the transformed environment. The hell on earth which was created during the early phase of the industrial revolution, for example, was not a necessary feature of industrialization as such. All of its technological and productive accomplishments could have been wrought - assuming that they were really necessary for human satisfaction - with far less damage to the natural environment and the living things therein (human and non-human), had the enterprise been differently motivated and regulated by a slower tempo. In actuality the motor of progress was and is oiled by the drives of greed, envy, and indifference to suffering. However potent the magic of the celebrated 'invisible hand' might be, an enterprise with a genetic makeup such as this could not have been expected to sire very attractive progeny.

The point I wish to make is a simple one: by exclusively directing our energies outward in the search for additional resources, in the course of which non-human nature becomes nothing but a means for the satisfaction of human needs, we are diverted from examining the character and objectives of our material demands. Instead of probing the features of the ensemble of satisfactions and dissatisfactions that exists at the present level of consumption, we push ahead in the expectation that a wider field of material resources will compensate for what we find lacking in the means now at our disposal. This is not, it seems to me, a reasonable expectation. There are good reasons for believing that a further quantitative expansion of the realm of commodities would merely shift the ensemble of satisfactions and dissatisfactions onto another plane.

In the high-intensity market setting nature tends to be regarded from an exclusively utilitarian perspective. To say this is not to suggest that some other perspective on nature could serve as a cure for our social ills. I have emphasized this aspect of the situation only because I believe that it may clarify the problem of human needs in this setting. The problem is that we are encouraged to ignore the confusions that exist in the relationship between our needs and the means by which we attempt to satisfy them, because our attention is fixated on ransacking the environment for more and better resources.

As we transform the natural environment exclusively into an arena for the satisfaction of needs, the results of our activity reflect this state of confusion. The environment we create is thus a kind of mirror in which the confusions can be detected. In an article entitled 'What's Wrong with Plastic Trees?' Martin Krieger argues that our wants can be conditioned 'so that the environments which people learn to use and want reflect environments that are likely to be available at low cost ... Much more can be done with plastic trees and the like to give most people the feeling that they are experiencing nature.'[40] What actually happens in this progressive simulation of experience is that there arises a deep-seated confusion about means and ends. The ends or objectives of the immensely complex productive apparatus that constructs material objects for simulated experiences become more and more obscure. What makes them necessary? Why has the unsimulated experience of nature, for example, become so rare and so scarce that the manufactured surrogates must be substituted for it? One reason may be that so much of the natural environment has been appropriated for actual or potential service in the production of goods.

Opportunities for activities such as what Krieger calls the experience of nature become scarce because those activities become more and more dependent upon the acquisition of commodities which are supposedly essential for the enjoyment of them. (Recall Staffan Linder's point about the 'increased goods intensity' of time.) This is apparent in the case of camping activities discussed earlier. Activities are mediated more intensively through the acquisition of goods; in other words, the enjoyment of activities is associated with ever-larger sets of commodities and appears to be impossible in their absence. The ensemble of satisfactions and dissatisfactions that is embodied in the consumption of goods is attached to the activities, such as the experience of nature. The elaborate combinations of activities and goods represent forms of satisfaction for needs. The relationship between individual needs and these complex forms of satisfaction is ambiguous. This ambiguity is an important factor in the expanding search for new sources of satisfaction, but the intensified interplay between needs and commodities also intensifies the ambiguity. The process feeds upon itself.

Laurence Tribe has suggested that plastic trees and the other artifacts of a simulated environment 'represent nature abstracted to pure categories of human need.'[41] The artifacts toward which needs are oriented become more simplified - that is, less rich in their particular qualities and their subtle differentiating features - in relation to a similar process of simplification in the qualitative (as opposed to the quantitative) dimensions of needing. In one sense plastic trees are an expression of our desire to be unencumbered by the claims of other living things in our environment, since in homes and offices they require no care and along streets they are unaffected by air pollution. In another sense this wish to simplify our environment represents a state of indifference toward the character and objectives of our own desires. We need so many things that we cannot afford to devote much care and attention to the particular qualities of any of them - or to the qualities of the needs to which they are joined. Having been pumped dry by the incessant solicitations of the marketplace, the wellsprings of need dessicate. The fragmentation of needs and their orientation to ever-larger sets of commodities fosters a sense of indifference toward each particular need, and thus toward the objects that are supposed to satisfy them.

In our orientation to non-human nature as only a support system for human wants - a warehouse of resources and a dumping-ground for wastes - we find the ecological context of our production and consumption activity that is characteristic of a high-intensity market setting. In each dimension of this activity - the individual, the social, and the relationship to the natural environment - we can detect different elements of the ambiguity and confusion that arise in the dynamic connection between needs and commodities in this setting.

PART TWO: DIAGNOSIS

Needs

Any attempt to understand what is meant by the experience that we call the satisfaction of needs is a risky endeavour. The concepts of need and satisfaction are crude indicators for very complex activities. How can one hope to correlate general patterns of overt behaviour, such as the purchase of commodities, with the covert psychological responses to the anticipated satisfactions that are at least to some extent unique for each individual? In 1890 the economist Alfred Marshall wrote: 'It cannot be too much insisted that to measure directly, or *per se*, either desires or the satisfaction which results from their fulfillment is impossible, if not inconceivable.'[1] Recently consumer researchers, employing devices such as the salivation tester referred to earlier, have been trying to correlate product messages and involuntary physical reactions, but as yet there are no indications that they will be able to scale the intensity of satisfaction. The mere attempt to do so, I suspect, would arouse great anxieties in many persons, irrespective of its chances of success.

In the foregoing remarks I have referred to an 'ensemble of satisfactions and dissatisfactions' that may be presumed to exist in the consumption experiences of individuals in the high-intensity market setting. In other words, we may assume that individuals derive in varying degrees some kinds of satisfactions from consumption, without trying to specify the nature or qualities of those satisfactions. We may also assume that there exists simultaneously a set of dissatisfactions. Messages about new products, for example, attempt to arouse feelings of dissatisfaction with existing products, and today individuals are continually bombarded with such messages. In short, there is a complex network of what we might call positive and negative feelings, constantly forming, dissolving, and re-forming, that constitutes the dynamic bond between the individual's perceptions of his needs and his judgments about what can satisfy those needs.

The idea that there is a network of positive and negative feelings involved in consumption activities has not been tested in empirical research. A study in a related area, however, lends some plausibility to it. That study attempted to measure people's perceptions of their happiness and well-being in terms of factors such as job and marital satisfaction. The researchers discovered that there was not a unified feeling of happiness or unhappiness but rather 'two distinct and independent dimensions' - positive and negative - that correlate separately with happiness or unhappiness but do not correlate with each other: 'Forces contributing toward increased negative feelings, such as anxiety, marital tension, and job dissatisfaction, do not produce any concomitant decrease in positive feelings, and those forces which contribute toward the development of positive feelings, such as social interaction and active participation in the environment, do not in any way lessen negative feelings. Thus it is possible for a person who has many negative feelings to be happy, if he also has compensatory positive feelings.'[2] It is impossible to determine whether this bifurcation of experience into two autonomous domains is a recent development or a continuation of an older pattern. In any case the description of the individual's experience as a consumer that has been given in the preceding sections is consistent with the findings of this research. The fact that people apparently separate feelings of satisfaction and dissatisfaction means that market messages can play upon both simultaneously. The ante in the game of consumption, so to speak, can be raised steadily, and the players will continue to bid, in part because they simply cannot calculate the net result of their gains and losses. This may explain why many Americans report in opinion surveys that they think earlier generations were happier than they are, while at the same time they maintain that they would not wish to return to the situation of those periods.[3]

One of the major presuppositions of my argument in this essay is that the high-consumption ideal tends to orient all aspects of an individual's striving for personal satisfaction toward the realm of commodities. Satisfaction is an elusive and idiosyncratic state of feeling, an ultimately unfathomable domain full of ambiguities and paradoxes; and in this setting enormous social pressures are brought to bear upon that domain. The ambiguities are intensified as the ante is raised, i.e., as the conviction grows that the best approach to the problem of satisfying our needs consists in fostering uninterrupted economic growth.

I have suggested that this is an unwise choice for a society even if quantitative material growth in the economy can be sustained indefinitely. The reason is that the tendency to define the pursuit of well-being exclusively in these terms involves increasingly grave risks for the individual's integrity as a person and for our relationship to the natural ecosystems that sustain our activities. The most common ways of understanding our needs that are presented in contemporary social theory, however, do not allow us to grasp the nature of those risks with sufficient clarity. In what follows I have tried to lay the foundations for a more adequate theory. By analyzing the process of needing in relation to specific, socially organized modes of need-satisfaction, I hope to justify my contention that our perceptions of our needs become progressively more confused under current circumstances. Only the scaffold for a new project has been erected in these remarks, and the scaffolding must undergo further testing before its usefulness as a means of addressing the problem of our needs can be assessed.

This essay focuses specifically on the interplay between needs and commodities in a high-intensity market setting. Of course both needs and commodities are factors in many other social settings. The act of focusing is more successful when one first surveys the horizon of the entire problem-field, for the landmarks that one seeks may be scattered among thick clusters of underbrush. This essay does not attempt to map in detail the entire region, but only to determine what aspects of the process of needing and what attributes of present-day commodities are relevant to the problematical character of satisfaction in the contemporary market setting.

What are the general characteristics of the region of needs and satisfactions in human society? We are innately social animals with an extraordinary capacity for developing those symbolic representations that are summed up in the term *culture*. In the intensive social interactions that have been always present in human societies, men and women collectively interpret, reflect upon, and integrate those otherwise spontaneous impulses that we normally call the striving for self-preservation in a species. Impulses are mediated or transformed by cultural forms into needs, that is, conscious expressions of desires that become 'congealed' in socialization patterns transmitted from one generation to another. The cultural, symbolic, or 'reflective' transformation of impulses breaks the direct interaction between instinctual drives and the means for gratifying them that normally occurs for other species through adaptation to environmental fluctuations. The development of tools is one of the most common expressions of the reflective transformation of the impulses for gratification, for this activity is dependent upon the ability to employ some aspects of the environment's resources as instruments so as to increase the overall range of possible sources of satisfaction. The direct interaction between impulses and sources of satisfaction is broken; impulses are controlled and consciously directed toward an enlarged field of satisfaction.

One of the outstanding characteristics of human development is this movement away from reliance upon the spontaneous providence of nature for the satisfaction of needs. In other words, there is an increasing intervention in and management of environmental processes, inaugurating a reciprocal interplay between symbolically mediated impulses (needs) and organized techniques for provisioning them. In the highly developed stages of this process the division of labour emerges; the individuals engaged in specialized production cannot satisfy all their needs directly but must depend upon regular exchanges of goods in the marketplace. For long periods and in many types of societies the scope of market exchanges and the division of labour does not develop beyond a certain point, since this would be inconsistent with other customs and institutional arrangements. (Illustrations of this point are given later in the section on commodities.) The different social functions of exchanges and markets in pre-modern societies have been recorded in anthropological and historical studies.

The high-intensity market setting is a generalized or unlimited market economy wherein the tempo of exchange for goods and services - constituting the interaction between needs and commodities - increases steadily both in scope and in the rate of change. A better understanding of its specific features is necessary because of the inadequacy of existing approaches. Theories of needs generally ignore the mundane sphere of everyday life in which individuals struggle with the interpretation of their needs, and theories of consumer behaviour usually avoid any critical evaluation of the suitability of commodities for our needs. An adequate understanding of this setting can be achieved only if each of the two interacting components is examined in relation to the other.

In the foregoing I have used the terms *wants, needs*, and *desires* interchangeably. This was done deliberately, so as to remain faithful to the terminological differences that exist in the sources which I have been examining. By bringing together contributions to the theory of needs from such fields as anthropology, economics, and psychology, one can perceive the role that terminological differences themselves play in inhibiting a critical study of the problem of satisfaction. The differences cannot be classified along strictly disciplinary lines, but there are some general patterns to be found in the contemporary literature. Sociology and humanistic social psychology often employ the word *needs*; economics uses *wants, choices*, and *demands*; behavioural psychology prefers *drives*; consumer psychology likes *choices* and *desires*. Finally, some philosophers distinguish all of these from *preferences*. Presumably it is the same person or social group that exhibits needs, wants, demands, drives, choices, desires, and preferences; but the variations in the concepts that are used make it extremely difficult to develop an integrated perspective on the concrete social situation of the high-intensity market setting.

Moreover, in most studies a basic dichotomy is established. The entire ensemble of needs or wants (or any of the other designations) is grouped according to the opposed categories of natural vs. cultivated, immediate vs. derived, viscerogenic vs. psychogenic, necessary vs. luxury, physical vs. symbolic, true vs. false, objective vs. subjective, basic vs. discretionary, and lower vs. higher. And then (to complete the confusion) there are various tables, hierarchies, and classifications wherein needs or wants are ranked according to some scheme of relative importance.

The attempt to classify needs into these kinds of categories began with Plato's *Republic* and continues in many present-day studies. A critical review of this literature would require a lengthy volume in itself and therefore cannot be undertaken here. Some attention must be paid, however, to the key issues that are raised in studies on the classification of needs, so that we may determine what perspective on the problem of human needs is relevant to the understanding of contemporary society.

THREE PATTERNS OF THINKING ABOUT NEEDS

There are three recurring patterns in the diverse terminological usages and classificatory schemes referred to above: the distinction between biological and cultural components, the matter of hierarchies or relative priorities, and the difference between behaviourist and critical perspectives.

The oldest and most common framework that has been developed for understanding the structure of human needs rests on the attempt to distinguish between a biological dimension and a cultural or social dimension. This is usually expressed in a listing of 'basic human needs' which begins with the physiological requirements of the individual organism, such as the drives for food, water, air, warmth, and sexual activity (a species need); occasionally a rudimentary 'psychological' dimension is added at this level, consisting of factors such as role-specifications that maintain a degree of group harmony among social animals such as homo sapiens. All other needs are seen as 'derivations' from cultural formations. The difficulty with this approach is that it is so abstract. The symbolizing or cultural activity of human beings is so intense and so complex that the biological-cultural dichotomy is never *present* in the everyday activities of social groups, whether they are primitive societies or industrialized empires.

Take the case of food. What could be more basic than the drive for nourishment? Yet the human activities of selecting and preparing (and avoiding) the natural substances that can serve as the basis of nourishment are embedded in the most incredibly intricate set of culturally-determined practices; for example, Lévi-Strauss' *The Raw and the Cooked* provides an elaborate illustration of this point in connection with his exploration of the myths of certain South American tribes. The available anthropological record reveals the great diversity in all the human practices devised in response to physiological requirements, including the instances in which certain individuals will commit suicide by depriving themselves of survival necessities so as to maintain the integrity of the social group. The evidence on this point seems to support Dorothy Lee's contention that 'it is value, not a series of needs, which is at the basis of human behavior.'[4] According to Lee, needs themselves are derivative, not basic. They are not the underlying foundations which explain the orientation of individual behaviour, but rather are themselves derived from a more fundamental set of values, varying from culture to culture, whereby social cohesion is maintained through an authoritative socialization process which governs an individual's interpretation of his needs. So tightly interwoven are the biological and cultural dimensions that, when they are separated into distinct categories, the result does not do justice to the concrete unity of human behaviour in practical, everyday situations.

The anthropologist Bronislaw Malinowski noted that 'this cultural, roundabout way of indirect satisfaction imposes secondary or derived needs.'[5] The main point is well taken: all human needs are routed indirectly toward sources of satisfaction by the agency of culture; this process does not give rise to 'secondary' needs, however, but rather blurs any attempted distinction between primary and derived needs. The biological-cultural distinction is purely abstract and cannot really help us to understand the diverse practical orientations of human behaviour, because every culture assimilates the biological dimension of existence into its socialization patterns, which are the actual determinants of the interpretations of need. The specific modes of response to socially interpreted impulses (which we usually call the satisfaction of needs), and not the formation of 'secondary' or derived needs, themselves represent a derivative activity in relation to the basic modes of social cohesion.

The deficiencies in the notion of basic human needs and in the biological-cultural distinction that is often associated with it can be seen most clearly when the following considerations are taken into account. All living things have requirements for nutrients, for example, and all social animals have physiological requirements not very different from the ones usually listed as basic human needs. What then does this argument tell us about basic *human* needs? These are simply the necessary preconditions for the maintenance of many kinds of animal life in general. Furthermore, if we look at the practices of other social animals, we find that many (if not all) of the so-called higher needs of human beings - such as affection, learning, co-operation, security - are also evident there. Their 'sociability' traits, transmitted intergenerationally through intensive socialization experiences, are simply the necessary preconditions for the existence of all social animals - for lions and apes as well as for mankind. Recent ethological studies demonstrate that the different members of the primary groups in such species develop well-defined individual characteristics, that they engage in elaborate role-playing, and so forth. In short, there is nothing uniquely human about either the physiological or the social-psychological requirements that usually make up the list of basic human needs.

Of course I am not suggesting that there are no differences between human needs and the needs of other social animals. The point I wish to make here is that the usual formulations of the biological-cultural distinction and the conception of basic human needs have not provided much insight into the unique characteristics of human need.[6] What really differentiates human beings and other social animals in this respect is the rich complexity of the human experience of needing, the fascinating diversity in the actual cultural expressions of need, that may be observed in the anthropological records.

The second major pattern of thinking is the arrangement of human needs into a hierarchy of ranks. Needs are classified into various groups which are said to have differing degrees of urgency and significance that determine the order in which we must attend to them. Again there are many examples of this pattern; in recent times the most influential has been Abraham Maslow's hierarchy of needs.[7] Maslow established the following categories of needs: physical (i.e. physiological); safety (order, predictability, dependability of the environment); love, affection, belongingness; self-esteem; self-actualization. These rankings proceed from the more immediate to the less immediate, so that the needs represented in each successive category appear only as the earlier ones are met to an adequate degree. The first three orders are basic or 'deficiency' needs, in that the absence of satisfaction for them results in physiological or psychological illness; the other two represent derived needs whose unfolding is the way to a fully developed personality. Release of tension is sought in attempts to satisfy the former, whereas the latter are open-ended positive activities. There is also an internal impetus in the succession, for it is only the latter orders which define the uniquely human condition and which are therefore the proper goal of individual striving. An ethical judgment and a conception of psychological health are incorporated in this scheme also. It is improper to remain fixated at the more immediate levels, for the measure of growth or maturity for an individual is the unfolding of the capacity to express the higher needs. For a mature person the 'material' objectives contained in the first two categories should find a satiation point where the accumulation of things ends, whereas the non-material objectives (the latter three) open up a vista of potentially endless personal growth.

What is apparent at first glance is that Maslow has adapted the biological-cultural distinction in conceptualizing a succession of levels that is both temporal and ontological in character. Biological needs are the most urgent and must be attended to first, and the others follow suit if society is properly organized so as to permit this to occur. The succession is at the same time ontological; that is, the later stages represent higher states of being, or the more appropriate domains of human existence as such. Neither the idea of temporal succession nor that of ontological development is convincing, however. In the first place, Maslow's arrangement reflects the specific organization of life in a technologically advanced society, where there is a high degree of specialization among functions and activities. This classification of needs seems far less applicable to the general pattern of earlier societies; activities that could be sorted into all five categories can be located there, but they are closely interwoven with one another and in my opinion they do not emerge serially as a succession of discrete individual or group requirements.

Is Maslow's scheme at least helpful in understanding the development of human behaviour in technologically advanced societies? In this context extensive market exchanges involving a great variety of commodities provide the organized framework for the satisfaction of needs. Maslow suggests that this level of activity should be transcended by each individual at a certain point in order to arrive at the level of self-actualizing activities, which are not related to the exchange of goods. But the actual behavioural tendencies in high-intensity market settings are far different. The sphere of material exchange is not transcended, but rather is extended ever more deeply into the 'psychological' domains. The needs for self-esteem and self-actualization are expressed and pursued through the purchase of commodities, which are not simply material objects but things that have a complex set of meanings or 'messages' associated with them. In this social setting there is little inducement for individuals to transcend their fixation on the world of objects – indeed, this would be directly contrary to the innermost tendency of that social practice which systematically orients personal activities toward the acquisition of ever-larger sets of commodities.

The way in which expressions of self-esteem and self-actualization function as integral components of a high-consumption lifestyle was depicted brilliantly more than half a century ago in Thorstein Veblen's *Theory of the Leisure Class*. Social developments since that time have merely intensified the process and have allowed a much larger proportion of the population in the industrialized nations to participate in it. The predominant message about the entire world of commodities as a whole in this context – the unifying theme in all the many messages about different things – is that self-actualization or 'individuality' *is* the goal of consumption, and that individuality is attained by assembling a unique collection of commodities.

From Maslow's perspective one might object that this is a perverted form of self-actualization. But how could one justify such a claim? According to Maslow self-actualization encompasses 'non-striving, non-self-centred, self-validating end experiences'; in specific terms it is reached in 'peak experiences' such as creative moments, intellectual insight, parental feeling, mystic intuitions, and so forth. This conception rigidly separates the non-material from the material sphere of life and implicitly denigrates the latter. But if many individuals interpret their peak experiences in relation to consumption, if these 'refined' feelings are embodied in commodities, all the Sunday sermons about the ultimate priority of spiritual matters and all the writings of sensi-

...ologists will remain on the periphery of the dominant social ...rns. One simply cannot divide the non-material and the material dimensions of needs. In industrialized as well as in other societies the ensemble of needs constitutes a uniform sphere of activity, each segment of which mirrors the common characteristics of the whole.

The question of whether there are distorted forms of need expression and fulfilment in a culture as a whole brings us to the third pattern of thinking about needs. Different answers to the question are given by those who adopt a behaviourist perspective (such as George Katona) as opposed to a critical perspective (developed by writers such as Erich Fromm). Using the concepts of demands, drives, and choices, the behaviourist analysis offers a descriptive account of observable actions (consumer purchases, for example) that includes no judgment about the intrinsic quality or appropriateness of those actions. The critical viewpoint, on the other hand, seeks to establish some objective standard in relation to which the appropriateness of actions can be judged. For example, Erich Fromm has argued from this perspective that 'a sane society is that which corresponds to the needs of man - not necessarily to what he *feels* to be his needs, because even the most pathological aims can be felt subjectively as that which the person wants most; but to what his needs are *objectively*, as they can be ascertained by the study of man.'[8]

Perhaps the clearest illustration of the difference between these two perspectives is their evaluation of the possible distinction between 'real' and 'manipulated' needs. According to the proponents of the critical viewpoint, psychological as well as physiological health is an objective condition, and therefore we can specify what is genuinely needed (and contrariwise what is really harmful) for individuals. This perspective has been used, of course, to criticize the manipulative tendencies of contemporary capitalist societies, wherein (it is alleged) 'artificial' wants or 'false' needs are created by social pressures, especially the constant barrage of advertising. From a behaviourist viewpoint, however, this argument is based upon an invidious distinction. All wants arise out of social conditioning; this process occurs in every culture, although its specific forms vary considerably; through social conditioning individuals learn to interpret their needs and to adjust themselves to the prevailing modes of approved behaviour.[9] In individual cases the social conditioning process may go awry (for example, obesity resulting from over-consumption of food); but so long as most persons continue to function within accepted parameters of normalcy, we have no grounds for the claim that the process is inculcating false needs.

The critical viewpoint has based its negative judgment of the high-consumption lifestyle on some form of the distinction between spontaneous vs. artificial or true vs. false needs. The sheer volume of advertising in modern capitalist societies appears to lend this thesis a plausible air. Why else would such an intensive effort at persuasion be necessary? Yet if the socialization process is so intense that the imperatives of the capitalist market economy itself (the necessity in the productive system for continually expanding the realm of commodities) become internalized as deeply felt needs in the experience of individuals, as Marcuse has argued, are there reasonable grounds for describing them as false needs?[10] This argument, far from gainsaying the behaviourist hypothesis, tends to reinforce it. Admittedly one may then argue that 'the whole is false,' i.e., that the entire system of needs thwarts the development of autonomous, self-actualizing individuals who are capable of forming non-exploitative interpersonal relationships. So long as the society as a whole and the majority of persons in it continue to function effectively enough to perpetuate the 'false' system, however, any proposed alternative remains merely an abstract possibility. It may well be an inherently self-destructive system, but only its general collapse would verify the hypothesis.

The principal difficulty in the critical perspective is the generally unconvincing mien of its contention that (in Fromm's words) the 'study of man' will supply an objective standard for judging the relative authenticity of felt needs. The proponents of this thesis are well aware that traditional, authoritarian moral philosophies - especially the religious variants - also postulated an 'objective' framework for evaluating human desires. They argue that the outlooks of these philosophies were conditioned by the specific historical circumstances in which they developed, and they also readily concede that there is no universally applicable objective standard. These failings notwithstanding, they maintain that an evaluative framework that is grounded in a critical study of modern social development indicates a way toward a different and better determination of needs by individuals in a future, non-authoritarian form of society. The conundrum for the theory is the difficulty in depicting realistically the stages through which persons who have been conditioned by heterogeneous needs - the consumer demands which actually reflect the market economy's own requirements for continuous expansion - could proceed to the state of autonomy and freedom.

Theoretical difficulties are not confined to this perspective alone, however. The strength of the critical viewpoint consists in its detailed examination of the conditioning process itself. For wants are not socially conditioned *in vacuo*, but rather are shaped in historically specific forms which are the outcome of differential class interests and power relationships. Merely to state that all wants are socially conditioned tells us nothing. The rat's behaviour in an artificial environment may be highly manipulable, but the environment itself does not appear spontaneously; in part the result is predetermined by the interests of the researcher who controls the set of variables. In the formation of human needs the environmental variables are embodied in the particular socialization process of each culture. If the study of the social conditioning of wants does not include a critical examination of the conflicting pressures which are brought to bear upon each individual in the socialization process, its approach will not yield very significant results. Especially in a society such as our own, with its emphasis on individual responsibility, the dramatic aspect of this process is the individual's struggle to form his sense of identity in the context of the powerful social pressures emanating from the market economy which extol models of conformity. The valuable contribution of the critical perspective is its insistence that the formation of needs can only be understood as the outcome of a dynamic interaction between individual psychology and socio-economic interest.

With the exception of what I have referred to as the critical perspective, the approaches to the study of human needs adopted in these three patterns of thinking are not very helpful for evaluating the problem of satisfaction in the high-intensity market setting. I am not suggesting that they make no contribution at all to the understanding of needs, but only that they are not helpful in the context of the problem discussed in this essay. The reason for this is that in attempting to achieve a high level of generality these theories develop fixed and abstract categories; the distinctions represented in these categories - such as those which make up the hierarchy of needs - tend to dissolve when they are applied to actual social situations. As I noted earlier, these categories are so broad that they may be applied to the behaviour of all social animals to some extent, and therefore they do not tell us very much about the specific character of *human* needs. The rich diversity of ways in which human beings interpret their needs is in my view the chief distinguishing feature of human activity in relation to the activities of other social animals. And the formation of human needs in specific settings can be understood best as a function of social patterns or frameworks of satisfaction. (There are four such frameworks, broadly speaking, which will be discussed in the section on commodities.)

The approach to the problem of needs that I have adopted in this essay is a modification of the critical perspective discussed above. In this perspective a distinction has been made between true and false needs. This has also been expressed as a distinction between wants and needs: wants express a person's subjective desires (his felt requirements), whereas his needs are his objective or true requirements. Throughout Part One of this essay I have referred to a 'confused' state of needs – a formulation that seems to me preferable to the distinction between needs and wants or between true and false needs.

NEEDS OR WANTS?

Is it possible to draw a line of demarcation between needs and wants? Can one say that an individual's needs are those minimum requirements he shares with all others in his own culture which must be met for the sake of their physical and mental health, and that his wants are his subjective and perhaps arbitrary desires that reflect his personal idiosyncrasies? And is it the case that needs are limited, whereas wants are unlimited?[11]

Different answers are given to these questions in part because of the different ways in which the terms are defined. The basis of argument for some of those who accept the distinction is the contention that needing is an objective state of being and wanting is a subjective (or purely psychological) feeling. The various lists of basic human needs, beginning with physiological requirements and gradually extending to sociability needs, are drawn up to support the contention; everything that is left over, so to speak, is relegated to the inferior status of a want. Since wants arise in our inner states of feeling, presumably we always know what we want; but since needs are objective requirements, it is possible that we may not know what we need.

The physiological requirements for food, water, shelter, and so forth are invariably referred to as the most obvious dimension of needs. Here at least a demarcation between needs and wants should be feasible. But once again the level of abstraction makes the argument trivial. What the individual organism objectively requires is a minimum nutrient intake, proper conditions for retaining or dissipating bodily heat, and socialization experiences to maintain group cohesion in social animals such as man. These are everyone's 'existence needs.' But such needs can be satisfied under a great variety of circumstances, many of which would be considered abhorrent by most persons today. Indeed these needs can be satisfied most efficiently in a setting where the environment has been ruthlessly simplified and organized for just that purpose. Such a setting is described in Zamyatin's famous dystopian novel *We*, where everyone is assured the necessary nutrients and shelter to sustain life. The sole nutrient is a bland petroleum derivative, however, and the shelter is a small glass-walled cubicle furnished identically for all. The 'needs' of all are thereby 'satisfied,' and the lifelong conditioning process in Zamyatin's imaginary society ensures that no wants that might transcend the sphere of needs will emerge.

The illustration may seem unfair, a simple *reductio ad absurdum*. I do not think that it is, because it does show that the real problems about the satisfaction of human needs arise only when we abandon the abstract categories of food, clothing, and shelter, and the similarly abstract characterizations of sociability needs (security, self-esteem, and so forth). All the most interesting and important issues arise when we study how the objective necessities of human existence are filtered through the symbolic processes of culture and of individual perceptions. In short, all the most important issues arise just in that nebulous zone where the so-called objective and subjective dimensions meet. It is trivial to calculate the need for food in terms of minimum nutritional requirements, for example. The real issues are: What kinds of foods? In what forms? With what *qualities*? And how does the perceived need for certain kinds of foods stand in relation to other perceived needs?

If we attempt to answer these questions, the distinction between needs as objective requirements and wants as subjective states of feeling breaks down. It is important to understand this point because what is at stake is not only a theoretical or conceptual approach, but also practical issues of social policy. What is detrimental about the attempted demarcation between needs and wants is that it encourages us to regard the sphere of needs largely as a quantitative problem: each person needs a certain amount of nutrients, shelter, space, and social services. The practical outcome of this statement of basic needs is reflected in some of the social policies of the existing welfare state: bulk foodstuffs for the poor, the drab uniformity of public housing projects, and the stereotyped responses of bureaucracies. The qualitative aspects of needs are suppressed in these policies, just as the qualitative aspects of needs are suppressed for society's more fortunate members in the quantitative expansion of the realm of commodities.

The distinction between needs and wants tends to detract attention from the qualitative dimension of the problem of needs. I think it is precisely the suppression of the qualitative dimension of needs which represents the most problematic aspect of the high-intensity market setting. Rather than attempting to understand this setting by distinguishing between needs and wants, therefore, I have proposed that we look upon this setting as one in which individuals become increasingly confused about the relationship between their needs and the means through which they try to satisfy them.

In what follows I will show how and why this confusion arises out of the ambiguous character of both needs and commodities. I do not claim that in any other social settings there has been or can be some kind of perfect clarity about human needs and that the present state of confusion may be contrasted with an ideal condition of unambiguous satisfaction. My purpose is to isolate the specific sources of the confused state of needs in one particular setting and the specific kinds of personal, social, and environmental risks that arise in this situation.

The state of confusion is rooted in the interplay between needs and commodities. This is not a matter of false needs. In my view the individuals who search for the satisfaction of their needs in the jungle of commodities have for the most part a reasonable and sensible set of needs; however, they do misinterpret the relationship between their perceived needs and the possible sources of satisfaction for them. Of course there are examples of unreasonable and even pathological expressions of individual needs, but these are not a typical feature of the behaviour in question here, namely the everyday activity of the ordinary consumer.

THE DUALISTIC NATURE OF NEEDING

In reviewing the theories of needs most commonly found in contemporary studies, I argued that for the purposes of this essay one could not accept the distinction between biological and cultural components in the formation of needs. Second, I suggested that one could regard the 'system of needs' as a continuous series of interactions among different levels of significance rather than as a hierarchical order. Third, I noted that one should not attempt to fashion a list of authentic needs for our society or to distinguish between needs and wants.

To understand the problem of satisfaction in the high-intensity market setting we require only one hypothesis about the structure of human needs. The hypothesis is quite a simple one, namely that every expression or state of needing has simultaneously a material and a symbolic or cultural correlate. The term *correlate* seems to be the most appropriate concept in this context for the following reason: although the two aspects of needing can be distinguished for purposes of discussion, they never function separately in the actual process of needing; in other words, the experience of needing is inherently a multidimensional activity.

Every facet of human needing has a material correlate in the sense that the life-requirements of individuals and societies necessarily entail a constant 'material exchange' of organic and inorganic substances governed by the structure of nature. Even if one were to regard something like spiritual contemplation as among the needs of mankind, for example, one must recognize that this need could only be expressed in an intrinsic relationship to an entire 'system of needs' (Hegel's phrase) which would also include some form of the material reproduction of life. On the other hand, every facet of human needing has a symbolic correlate in the sense that the material exchanges of life are mediated (i.e., interpreted through the cultural or reflective transformations of impulses) by means of elaborate social interaction patterns. Neither the material nor the symbolic aspects can be reduced or collapsed into the other; nor in my opinion (contrary to arguments by writers such as Baudrillard) can the two be separated, so that one could claim to detect a 'symbolic order' that is divorced from the sphere of political economy. I believe that the human system of needs in every culture is an indissoluble unity of material and symbolic correlates.

A long-range perspective may help to clarify this point. In human development there is no 'direct' appropriation of nature, no simple or unidimensional correlation between internal and external stimuli on the one hand and the organism's activity on the other. The vast symbolizing capacity of the human brain, which operates within the context of continuous and intense social interaction, filters impulses through intricate mediations which govern the interrelationships among the individual, the group, and the natural environment. Human development is not a progression from relative simplicity to relative complexity in the articulation of needs. Anthropological researches demonstrate that an incredible degree of complexity in the expression of needs has existed in so-called primitive cultures, and we may safely infer that this is the outcome of a process that reaches back into the evolutionary origins of our species. The available evidence shows clearly that there is no aspect of our physiological requirements (the famous basic needs for food, shelter, and so forth) that has not always been firmly embedded in a rich tapestry of symbolic mediations. Likewise what are called the higher needs - love, esteem, the pursuit of knowledge and spiritual perfection - also arise within a holistic interpretation of needs and are not separated from the material aspects of existence.

The complexity of human needing is inherent in the species' evolutionary origins; it is not itself the result of historical or cultural development, but is rather the necessary presupposition for that process. There is no movement from simplicity to complexity, only a progressive unfolding and enrichment of the symbolic mediations subsisting in reciprocal relation with the continuous reinterpretation of needs.[12] The potential elaboration and variation of this innate symbolizing capacity was initially dependent upon the evolutionary success of homo sapiens vis-à-vis other hominid species and other mammals. The later transition from a migratory to a sedentary ecological pattern reflected a further adaptive response in accordance with the genetic makeup of homo sapiens. The refinement of desire and the environmental manipulations that are associated with the sedentary state opened new vistas for the instrumental applications of the symbolizing capacity in building, crafts, agriculture, and so forth.

The rich genetic endowment of homo sapiens boded ill for its natural competitors from the beginning, but one need not assume that there was any strict necessity in the process leading up to our present situation, where a single species organized into permanent settlements is so predominant on the planet. This situation represents one possible outcome in which contingent factors undoubtedly played a part. One can distinguish loosely the following types of human ecology, in all of which the duality in needing is present to some degree: the hunting-and-gathering ecology with an increasingly formalized socialization process; small-scale permanent settlements, initiating the production of the means of subsistence; large-scale settlements and then civilizations, with considerable spheres of exchange and the division of labour; capitalism and industrialization, in which the high-intensity market setting emerges after some time. This is not to assert, however, that the succession of different types of human ecology was uniform, linear, or necessary either within these types or in their relations to each other.

The emergence of production for exchange marks a profound deepening of both the intensive and extensive dimensions within the duality of material-symbolic correlates in human needing. Considered intensively the duality becomes far more subtle. Exchanges among different small-scale cultures create an interface between the holistic domains of symbolic mediations that are present in each, and in large-scale permanent settlements exchanges develop among members of the same culture who stand in relation to one another as possessors of specialized talents, none of which is sufficient for the complete satisfaction of their needs. The differences between the smaller and the larger cultures are a matter of degree rather than kind. In the common pattern which has been called the domestic mode of production, consisting primarily of production for use but including limited exchange relationships, the members of a social unit generally engage in the same tasks and produce similar things independently.[13] Greater specialization of labour and exchanges within and among societies increase the complexity and number of the intermediate steps required for the satisfaction of needs, and the dual material and symbolic correlates present in each step take on finer shadings and nuances.

Considered extensively, the expanding role of production for exchange bears a quite obvious connection with this dualistic character of human needing. Whether we prefer to speak in terms of the creation of new needs, the refinement of existing needs, or the fragmentation of needs, the increasing number and variety of goods which enter the realm of personal experience through exchange transactions are the physical vessels in which the dualism is now embodied. The mere quantitative expansion of the realm of things does not itself affect the complexity of the process; but the increase of goods tends to shift the locus of the cultural mediation of needs from the non-material domain (myths, legends, taboos) to the material domain (physical objects). Once again I must emphasize the point that the change is a matter of degree and not of kind, since the symbolizing activity in all cultures always refers in part to the physical sources of satisfaction in the natural environment.

The general tendency of the present-day market economy, impelled by industrialized mass production and much wider commodity exchange, is to embed the network of symbolic mediations that shape the character of human needing exclusively in material objects (or more precisely, to orient needs entirely toward commodities). This process is what Marx called the fetishism of commodities. In the production and exchange of goods in the market there is 'a definite social relation between men that assumes in their eyes the fantastic form of a relation between things.'[14] The phrasing in this famous passage is too strong, however. We must be careful not to draw too sharp a distinction between production for use and for exchange. In both cases the appropriation of nature is mediated through complex cultural forms; in other words, there is in both a dynamic interaction between the material and symbolic correlates of human needing. The focusing of needs in terms of this duality, which has varied so greatly among the world's cultures, permutes the shapes and proportions of things in the field of satisfaction; yet it is a single lens - the social activity in which both use and exchange occur - that transmits the kaleidoscopic images of desired objects. There are significant qualitative turns, but no clearly identifiable breaks, in the continuum of needing within the sedentary phase of human development that extends from the simplest subsistence economy to the modern industrialized society's meretricious assortment of goods.

Commodities are exchanged goods; there is a separation between the acts of producing and consuming them which occurs during their journey through the marketplace (in industrialized production this journey is a lengthy one). The material-symbolic duality oscillates in the gap between the activities of producers and consumers with respect to any ensemble of commodities; the larger the ensemble, the higher is the oscillatory frequency. Commodities are things that are intended to satisfy needs, and therefore the duality of material and symbolic correlates is embodied in them. Commodities incorporate this duality and the ambiguous character of needing that is associated with it. The element of ambiguity is intensified in a large-scale market economy on account of the wider separation of the activities of production and consumption. This point will be pursued more fully in the section on commodities.

THE ECOLOGICAL CONTEXT OF HUMAN NEEDING

In Part One three aspects of the problem of satisfaction were distinguished: those pertaining to the individual, to society, and to non-human nature. They can be isolated from each other only for purposes of analysis; in actual everyday life these three are integrated facets of our behaviour. Having followed the route marked out by them, we should now examine the path more closely in the light of the hypothesis about the ambiguity in human needing that is rooted in the duality of material and symbolic correlates.

Production and consumption in contemporary industrial societies have a significant impact on global ecological systems such as the earth's atmosphere. This means that the ecological context of our needs is the planetary biosphere as a whole. Considered as a living entity, the biosphere and its component elements are characterized by a striving toward homeostasis, which is a state of dynamic equilibrium wherein an organism responds continuously to external and internal stimuli and seeks to balance them within certain parameters.[15] On the planetary level, for example, this involves the interaction between external forces (radiant energy from outer space) and internal forces working in the earth's crust and core. On the level of living species, the homeostatic impulse is expressed in the mechanisms of adaptation and genetic mutation through which the species as a whole attempts to preserve its viability in relation to its environment. Likewise in each individual organism the balancing of internal and external stimuli must be maintained within given parameters (for example, body temperature fluctuations in mammals) to ensure continued survival. The striving toward homeostasis is the biological context within which an organism's needs are expressed.

Homeostatic processes operate in the context of the evolutionary life-cycles that govern the development of species and ecosystems. There are normal parameters for these life-cycles as well. In the complex array of interactions that take place in an ecological subsystem, lakes as well as mammalian species appear and disappear. In the system as a whole and in the interactions among its component elements there are constant fluctuations in relation to an 'ideal' equilibrium state.

The human ability to manipulate the environment broadened the parameters within which this particular species evolved. The prevailing homeostatic mechanisms governing the interaction between this species and its environment, for example the regulation of population size, thereby underwent a decisive transformation. Clearly the parameters that define the limits of the homeostatic fluctuations in the planetary biosphere are flexible enough to have accommodated the implications of this human transformation so far. We can assume that there are some theoretical limits to this process, although we are not yet in a position to specify them on the basis of existing scientific knowledge. For example, our activities might result in major climatic changes that would affect the viability of life for us and other animals.

The human transformation of the environment introduces a new factor into the established homeostatic mechanisms of the biosphere. The extremely broad parameters within which these mechanisms function have allowed the planetary biosphere to adjust to the consequences resulting therefrom. At the same time the human species as a whole prospers vis-à-vis its erstwhile competitors. But at some point the beneficiaries must take cognizance of these consequences and seek consciously to 'manage' them, lest the as yet undeterminable limits - the biosphere's capacity to tolerate the impact of the numbers and needs of this prodigious species - be exceeded. The underlying implication in this transformation is therefore the necessity for subjecting the human manipulative capacity to rational control.

Understanding and managing the relationship between humanity and the natural environment is now a 'need' of the species. This need is implicit from the beginning in our ability to manipulate the environment; in other words, it is a hidden requisite in that manipulative ability itself. Only the broadness of the biosphere's functional parameters has allowed us to postpone our recognition of this fact for so long. To postulate the existence of a need for the rational mastery of the relationship between humanity and nature is to open the way toward understanding the full dimensions of the ambiguous character of human needing.

Without exception all of the schemes proposed to date for classifying the components of human needs (sketched previously in this section) have overlooked this critical element. All of them have detailed the various requirements of the individual organism on the one hand and the presuppositions for social cohesion among individuals and social groups on the other. Basic human needs have been listed and hierarchically ordered from a wide range of perspectives, but the larger context of human needing - the unique character of its orientation toward the natural environment - has been consistently ignored. It is almost as if basic human needs had been conceived according to the model of interplanetary explorers subsisting in an extraterrestrial environment. Imagining the situation of a permanent human colony on the moon, one could inventory the inhabitants' needs for food, water, oxygen, shelter, and waste disposal in one category and their needs for socialization patterns to maintain group integrity and cohesion in another. Since all of the means for satisfying these needs would have to be imported from elsewhere, there would be no intrinsic connection between them and their natural environment.

Yet in its actual earthly setting the character of human needing developed in close relation to the transformation of the natural environment. The powerful symbolizing capacity of the species, through which its needs are variously interpreted and oriented toward objects of satisfaction, made possible the extensive manipulation of environmental processes. The refinement of needs proceeded in reciprocal relation to achievements in the realm of environmental control. To conceptualize human needs simply with reference to the individual and social dimensions of their formation, in abstraction from their grounding in an orientation to the environment that is unique among all living species, is to obscure one of their most significant aspects. All three dimensions - the relation to nature, the individual, and society - and not simply the latter two are interwoven in the duality of material and symbolic correlates in human needing.

The possible future crises that may arise from the impact of human activities on the planetary ecosystem have been concealed both by contingent factors (the comparatively low level of attained control over environmental processes until modern times) and the adaptive capacity of the biosphere's homeostatic parameters. The implicit requisite condition, namely the necessity of exercising rational control over our manipulative efforts, has only recently become apparent. Henceforth we must evaluate the character of human needs in relation to the needs of non-human nature. In a later section of this essay I shall explore this conception and its implications for future social development.

The production and consumption of commodities in the high-intensity market setting, which is sustained by the progress of the industrial system, has introduced the possibility that major environmental crises may occur in the future. We have made the satisfaction of our needs dependent upon the steady expansion of this activity, and therefore a critical examination of the interaction between needs and commodities is essential. I have suggested that an element of ambiguity is present in both components of this interactive process and that this ambiguity lies behind the ensemble of satisfactions and dissatisfactions experienced in everyday life. The element of ambiguity in the peculiar character of human needing has been explored. A second element of ambiguity exists in the commodities that are produced in this setting.

Commodities

Modern social thought has been fascinated by that part of social activity revolving around market exchanges of goods and services. One of its most famous representative works, Adam Smith's *Wealth of Nations*, suggested that the inclination to exchange might be grounded in the symbolizing capacity of the human species. Smith thought it 'probable' that the 'propensity to truck, barter, and exchange one thing for another' was 'the necessary consequence of the faculties of reason and speech ... It is common to all men, and to be found in no other race of animals, which seem to know neither this nor any other species of contracts.'[16] He apparently believed that human reason and speech function through the exchange of ideas and feelings and therefore that exchange was a universal feature of human behaviour (the exchange of goods being one of its many aspects).

Later anthropological researches documented the ubiquity of exchange in various cultural patterns but they also indicated how varied and highly structured were its manifestations. Eighteenth-century thinkers like Adam Smith assumed that the extent of exchange activities in any society depended primarily upon the extent of the division of labour, and their intention was to show that social progress would result from encouraging the further development of this syndrome. Given the state of existing ethnographical knowledge at the time, they could not have realized that limitations on exchange activities were also a function of cultural determinants such as kinship associations and role specifications. They did not distinguish market and non-market types of exchanges. Nor could they have appreciated the co-mingled emotions of desire and fear that often accompanied exchange transactions among different cultures, on account of which the commercial contacts might turn into violent encounters.[17]

In the high-intensity market setting exchange activity occurs at a furious pace. Those who live in this setting learn to search for the satisfaction of needs in the purchase of commodities. The extraordinary array of commodities in the wealthier industrialized nations, however, represents a quite recent phase in the evolution of the market economy. In order to understand the unique character of the problem of satisfaction that arises in this setting, one must view it in relation to other patterns of social organization that have been devised for the satisfaction of needs.

FOUR SETTINGS FOR THE SATISFACTION OF NEEDS

In a preceding section I suggested that most theories of needs are incomplete because they attempt to analyze the components of human needs in isolation from actual social patterns of satisfaction. What are the general patterns of satisfaction? In the range of human societies four major types are apparent: production solely for use; production primarily for use, but including limited exchanges, in small-scale social units; production primarily for use, but including limited exchanges, in larger social units; and production primarily for exchange. This is a schematic list of 'pure types' which is useful for purposes of discussion. It is important to remember that in reality such neat divisions do not exist, that there are many variations among the cultures that can be grouped into these classifications, and that their features shade into each other and blur the abstract divisions that are presented here. Notwithstanding these qualifications, the typology enables us to establish a better perspective on our own situation.

Production solely for use represents the common pattern of hunting-and-gathering societies which have no permanent settlements. Their requirements for mobility dictate a minimal accumulation of artifacts. The members of the social unit normally produce all the artifacts that are necessary for the provisioning of their needs. This situation changes only when such peoples come into contact with others that are more technologically advanced. When this occurs they may begin to trade for items which they cannot fashion for themselves.

In small-scale settled societies production for use is organized around the extended-family unit which generally provides most of what is required for subsistence. In addition, however, there is an elaborate network of exchanges within and among the cultures of this type. The exchanges are limited to certain well-defined classes or types of objects. The things that are produced by the domestic units fall into different categories - for example, what are called prestige goods and subsistence goods - and a set of rules defines the kinds of exchanges among them that are permitted or proscribed. Among the

Tiv people of Nigeria the classes are established as follows: '[There are] three categories of objects: subsistence goods, prestige goods (slaves, cattle, metal), and women. Within each category one object can be exchanged for another. Between the second and third categories certain principles of conversion make possible access to women on the basis of brass rods, but goods of the first category cannot be converted into those of the second - nor, and especially not, into the third. There is thus no money serving as common denominator between these three categories, and labor and land remain outside them.'[18] The proximity or invasion of more technologically advanced cultures can disrupt the established patterns of behaviour in these societies just as severely as in the case of the nomadic hunting-and-gathering peoples.

Large-scale permanent settlements - civilizations - constitute the third pattern. In all pre-modern civilizations, as well as in those which remained economically 'backward' until the twentieth century, the majority of the population continues to depend primarily upon production for use in the satisfaction of their needs. In this setting non-market exchanges (managed by a state bureaucracy) may be quite extensive; in addition, market exchanges occur regularly but are still quite limited in scope, although they extend across far wider regions of the globe. One of the principal features of the second pattern - the fact that land and labour are not freely exchanged in market transactions - also obtains in the third, albeit in different forms.

One of these civilizations, namely European feudalism, prepared the seedbed for the emergence of capitalism, industrialism, and unlimited market exchange. Here all the 'factors of production' - land, labour, and capital accumulation - were transformed into an open-ended series of market transactions that promoted technological innovation and the division of labour at an accelerating rate. For the first time a market *economy*, as opposed to market exchanges, came into being. Production for use gradually dwindled, being confined largely to the 'backward' sectors of the economy (such as farming), and the majority of the population became dependent upon the purchase of commodities for satisfying their needs. In the more advanced industrial societies the final stage in the transition from production primarily for use to production primarily for exchange - as a determining factor in the lives of most persons - occurred around the turn of the twentieth century.

COMMODITY PRODUCTION: USE-VALUE AND EXCHANGE-VALUE

Out of the fourth pattern of need-satisfaction arose the high-intensity market setting. In this setting there is an enormous array of material things and the associated services which help to make possible the use and enjoyment of them. Commodities themselves here become very complex objects: the ambiguities inherent in the material-symbolic correlates of human needing itself are replicated in these objects. They are not simply material things but 'material-symbolic entities' - that is, things which embody complex sets of messages and characteristics. These incorporated messages offer suggestions to the purchasers of commodities concerning the suitability of the commodities for the needs of individuals. The theory of commodities that evolved from the eighteenth to the twentieth centuries gradually achieved greater clarity about the workings of this process.

The main thrust of modern political economy, especially in its mature phase beginning in the second half of the eighteenth century, was to demonstrate the social advantages flowing from the division of labour, the applications of technology to production, and the enlarged scope of market exchanges of commodities. Its primary focus concerned the material aspect of human needs in relation to market exchanges; the symbolic aspect was by no means absent, but it went largely unrecognized. In fact it was implicit in the theory of value that was developed in order to explain the nature of commodities; buried at first in the concept of value, it became apparent only in the later stages of the theory's evolution.

The theory of commodity exchange was rooted in a fundamental distinction between 'value in use' and 'value in exchange,' to employ Adam Smith's phrasing. The former denoted the inherent capacity of a thing to minister to a want or desire; the latter, the relative market power of a thing vis-à-vis all others, i.e., on what terms it might be exchanged in the market either by bartering or (in an open-ended market setting) through the common medium of money. In discussing the 'psychocultural background' of the value theory developed in political economy Walter Weisskopf has attempted to explain why the initial distinction between use-value and exchange-value was not carried through in the theory's further elaboration: 'The dichotomy of value in use and value in exchange reflects the conflict between the precapitalist attitude and the economic value complex. By rejecting the concept of value in use, the classical economists declared themselves implicitly as partisans of the new mode of life and the economic value complex. The market economy required that the individual strive toward acquisition, accumulation, and riches, without limit, as ends in themselves. The use of wealth was originally not the main purpose of acquisition. Thus use-value could not play a predominant role in the determination of economic values. Value in exchange is the important concept in an economic order in which people are supposed to exert themselves to the utmost in order to accumulate capital. Need satisfaction is only an incidental aim.'[19] Not market exchange as such, but the universalized version of exchange (the market economy) undermines the earlier practice of operating within discontinuous realms of objects (subsistence goods and prestige goods, for example) and constitutes the decisive alteration in the cultural context of economic activity. But Weisskopf correctly argues that the notion of use-value was rooted in an epoch that had passed, for it presupposed the existence of objective standards according to which a thing's appropriateness for an individual's needs could be assessed. This is why the concept of use-value received less and less attention in the main tradition of modern economic thought during the nineteenth century.

Political economy - and later modern economics - remained content with the attitude that the expanding sphere of market exchanges stood in some indeterminate, unspecifiable relation to the needs of individuals. Its basic assumption is that individuals are the only proper judges of the suitability of commodities for their needs. But Marx refused to acknowledge that the concept of use-value could be summarily dismissed. In the introductory chapters to his *Contribution to the Critique of Political Economy*, *Grundrisse*, and *Capital*, he sought to analyze the unique character of a generalized exchange economy by examining the dialectical interplay of use-value and exchange-value and of production and consumption.[20] Marx saw that one could not just take for granted the existence of generalized market exchange, for it does not appear spontaneously in human society like a *deus ex machina* but rather is the result of a specific historical process: 'The dissolution of all products and activities into exchange values presupposes the dissolution of all fixed personal (historic) relations of dependence of ... the producers on one another ... Prices are old; exchange also; but the increasing determination of the former by costs of production, as well as the increasing dominance of the latter over all relations of production, only develop fully, and continue to develop ever more completely, in bourgeois society ... The reciprocal and all-sided dependence of individuals who are indifferent to one another forms their social connection. This social bond is expressed in *exchange value*, by means of which alone each individual's own activity or his product becomes an activity and a product for him ... Relations of personal dependence (entirely spontaneous at the outset) are the first social forms [i.e. slavery, serfdom], in which human productive capacity develops only to a slight extent and at isolated points. Personal independence founded on *objective* dependence [i.e. capitalism] is the second great form, in which a system of general social metabolism, of universal relations, of all-round needs and universal capacities is formed for the first time. Free individuality, based on the universal development of individuals and on their subordination of their communal, social productivity as their social wealth, is the third stage. The second stage creates the conditions for the third.'[21] The gradual lifting of institutional restrictions on the activities of human labour and the articulation of needs constitutes the necessary social framework for universal exchange.

The most interesting aspect of Marx's attempt to identify the historical prerequisites of generalized exchange and commodity production is his emphasis on the interrelated positive and negative aspects of the process. The selection quoted above is only one representative instance of his oft-repeated assertion that capitalism establishes the necessary preconditions for the possibilities of the 'free individuality' that can emerge in socialism and communism. Taking a longer perspective, he assigned to exchange activity itself a prominent role in this dialective of positive and negative forces. Exchange is the 'chief means' through which the individual emerges from the 'herd-like' character of the social group and relates back to it as an associated member of a free community of persons.[22] The hermetic quality of the isolated social unit is overcome through increasing exchanges and affiliations; the positive aspect of this long historical process is the creation of the social conditions in which individuality becomes possible. Its negative aspect is the growing intensity of alienation. In production for exchange human labour relates to its 'objective conditions' (natural resources and the social relations which control the forms of access to them) as circumstances which are not under its direct control. Capitalism, which universalizes the scope of market relations and completes the divorcing of the labourer from the means of production, thus gives rise to 'the most extreme form of alienation.'[23]

Marx thus sees the process of exchange as a crucial element in human history, opening possibilities for the formation of individuality while simultaneously intensifying the alienated relationship of the human agent to his own productive activity. Casting this conception in the terms used earlier, one could say that both positive and negative aspects (individuality and alienation) were also mediated through symbolic forms in the process of being shaped by social class divisions. A deficiency in Marx's analysis is that in some passages he drew too sharp a distinction between 'direct production' for use and production for exchange.[24] Or, to phrase the point more accurately, he emphasized the elements of discontinuity between the two modes of production without drawing attention to the equally significant elements of continuity. The available ethnographical evidence does not permit us to distinguish a phase other than that of the nomadic hunter-gatherer that is characterized by production exclusively for use. A more reasonable speculation is that exchange appeared coterminously with the emergence of small-scale settled societies. The mere physical isolation of certain groups may have been the deciding factor in cases where exchange is entirely absent, or there may also be a few examples of socialization patterns in which it was explicitly discouraged or forbidden.

Marx's view underplayed the significance of the symbolic mediations inherent in all forms of the relationship between human activity and the natural environment. This led him to attribute an unequivocal, intrinsic quality to the use-value of exchanged goods. He described use-value as constituting the 'body' of commodities and as being composed of two elements, matter (*Naturstoff*) and human labour; in market exchanges, he maintained, use-values represent the 'material depositories of exchange value.' What seems to be implied here is that every commodity has an objective character in relation to human wants, even though Marx himself acknowledged that the 'nature of such wants, whether for instance they spring from the stomach or from fancy, makes no difference.'[25] But the perception of the usefulness of things related to needs such as food and clothing is, as I have argued above, conditioned by cultural or symbolic mediations; if this is correct, then no commodity has an unequivocal, objective character that arises simply out of its material attributes alone.

Marx indeed recognized that the later stages of industrialized production made the connection between need and use-value ambiguous, for he noted that 'the production in enormous mass quantities which is posited with machinery destroys every connection of the product with the direct need of the producer, and hence with direct use value ... '[26] This formulation approximates more closely the interpretation of the interaction between needs and commodities offered in this essay. There is a qualitative turn introduced by industrialized production, to be sure. But the duality of material and symbolic correlates in human needing does not allow us to postulate the existence of direct (i.e., unmediated) needs at any stage in the sedentary phase of human development. Generalized exchange and industrial production do not create that duality. Rather, they open the way toward its most radical and problematic expression.

In Adam Smith's *Wealth of Nations* the terms *use-value* and *utility* are already used interchangeably. And whereas Marx preferred the former term because it suggested that even in a generalized market economy there usually remains a direct link between needs and material things, the very different implications of the latter term began to be clarified in the dominant tradition of modern economic thought. The outcome of this new approach was an explicit denial that utility or usefulness is an intrinsic property of commodities themselves. The English economist Jevons expressed the consensus that had been reached by the last quarter of the nineteenth century: 'In the first place, utility, though a quality of things, is *no inherent quality*. It is better described as *a circumstance of things* arising from their relation to man's requirements.'[27] Utility is described as a relation, not as a property of an object. As such it is a highly ambiguous term - or perhaps it is an accurate designation for a highly ambiguous form of social activity.

Why does Jevons say that utility is a quality, but not an inherent quality, of things? The answer must be that anything may have the attribute of usefulness attached to it, depending upon the circumstances. Anything that is desired has utility (to a diminishing degree and in relative proportion to other desired things), and if the supply is scarce in relation to demand it will also have a market price. It is not for the economist to pass judgment on the peculiarities of the desire or to comment on the mysterious bond between the desire and its intended object; only adepts in depth psychology may safely venture into these uncertain waters.[28] But subsequent developments showed that conceiving utility as a relation between persons and things still did not do full justice to the subtleties involved in the articulation and satisfaction of needs. The concept of utility still bore the implication that there is some kind of unequivocal character in the object itself and that the 'indeterminate' aspect of consumption - the variations in different individuals' perceptions of a commodity's utility - consists in the relation between persons and things. To be sure, this does explain in part how certain ambiguities in the process of need-satisfaction might arise, since all sorts of personal idiosyncrasies are involved in the relations between persons and things. However, the development of the theory thus far still does not permit us to detect the powerful element of ambiguity that is incorporated within the commodity itself.

COMMODITIES AS COLLECTIONS OF CHARACTERISTICS

A few years ago an American economist, Kelvin Lancaster, developed a theory of the commodity which represents a major breakthrough in the understanding of present consumption activity. He noted that the earlier marginal utility theorists regarded the ordering of an individual's preferences as a matter of the relationships between 'wants' (the properties of an individual's psychological state) and the properties of goods. These theorists did not assume that there was a straight one-to-one correspondence - what I have called a univocal relationship - between a person's wants and the kinds of goods he chose for satisfying them. But they did assume that each commodity - or, more precisely, each group of commodities such as food and clothing - had a particular set of qualities (or univocal character) which distinguished it from that of other groups.

This was not an unreasonable assumption to make at the time, for it was consistent with the nature of the nineteenth-century market economy. There most commodities were still relatively simple in terms of their material composition, and each group of commodities did have a set of particular characteristics which were quite different from those of other groups; in addition, their qualities could be assessed through the common-sense craft judgments of an average person. But the contemporary situation is far different. Today's consumer operates in a market economy 'in which any specific characteristic might be obtained from any good. Thus the consumer's universe is, in principle, the complete set of all possible varieties of all possible goods - an astronomical number of goods, and presumably of characteristics.'[29]

What does Lancaster mean by the 'characteristics' of commodities? These are the real, physical properties of a thing which can be inspected and evaluated according to established criteria of evidence: 'Every objective property of size, shape, performance is a potential characteristic. In principle, if we take an object, measure it in every possible dimension and in every aspect of performance, in every biological, chemical, and physical aspect, we have evaluated all its possible characteristics.'[30] Characteristics are, therefore, aspects of what used to be called the use-value of commodities. There is nothing new in this assertion; what is new in Lancaster's theory is his further argument that consumers have a direct or immediate interest in characteristics and an indirect or derived interest in goods. This follows from the complex makeup of commodities in a modern industrial economy. Rather than saying that people have an interest in food items *per se*, for example, we should say that they seek a bundle of characteristics such as nutritive content, convenience of preparation, packaging of portions, appearance and texture, and so forth. The characteristic of convenience, however, is also an attribute of types of clothing, household appliances, modes of travel, and other groups of commodities. Lancaster argues that individuals differ primarily in their reactions to various characteristics, or in other words that they order their preferences directly in relation to collections of characteristics and indirectly in relation to the goods that possess those characteristics. Regarding this situation from another angle, one would say that 'a producer is ultimately selling characteristics collections rather than goods.'[31]

In adopting this perspective, Lancaster concludes, we would 'view the relationship between people and things as at least a two-stage affair. It is composed of the relationship between things and their characteristics (objective and technical) and the relationship between characteristics and people (personal, involving individual preferences).'[32] This theory represents a great advance over the older conception that each want is a want for a particular commodity as such. In a high-intensity market setting commodities themselves become fragmented into sets of characteristics and lose their independent 'identity,' so to speak; this process stands in reciprocal relation to the fragmentation of needs.

For Lancaster the significance of this analysis is that it shows the necessity for making available to consumers more adequate and objective information about the characteristics of commodities. In the case of automobiles, for example, the average individual cannot undertake his own evaluation of this commodity's performance characteristics; therefore, the kinds of test data provided in *Consumer Reports* studies should be made more accessible to consumers, if necessary by new legal stipulations for disclosure and verification of performance claims.

Lancaster suggests that more stringent legal requirements for contents labelling on containers would also increase the consumer's knowledge about a product's characteristics. Undoubtedly this would be an improvement over existing conditions. But would the full listing of polysyllabic chemical compounds on a container, for example, really improve the quality of an individual's judgments about the suitability of a product for his needs? As I have suggested earlier, in the context of the present vast array of commodities whose material composition is so complex, a simple increase in the amount of reliable information about them would not affect significantly the state of confusion about needs and the sources of satisfaction for them.

Perhaps this point can be illustrated by means of an imaginary situation or 'thought-experiment.' Let us assume that, in the department store section where the great numbers of personal 'health and beauty aids' are displayed, all the goods are packaged identically, are labelled only with actual product information (i.e., the listing of their chemical ingredients), and do not carry any brand names, graphic illustrations, suggestions for use, or invidious comparisons with competing items. Let us further assume that the prospective purchaser was born, raised, and educated in this society but has just returned to normal circulation after a twenty-year residence in an isolated mental institution. All the product information it would be possible to assemble would be to little avail under these circumstances, and the sheer number of objects for which there was no self-evident use would engender in him a state of utter perplexity. In other words a regular, sustained effort aimed at keeping acquainted with the *imputed* characteristics of things is required of the consumer in order to 'release' those objective characteristics embedded in the present array of commodities and to set in motion the interplay between persons and things.

Lancaster's theory can be modified so as to explain this point more fully. The first stage of his two-stage process (the relation between things and characteristics) is at least potentially a matter of objective information; the second - the relation between characteristics and people - is of course a subjective, psychological matter. When this two-stage process is reformulated as a statement about the nature of the commodity itself, the commodity appears as a complex unity of objective and imputed characteristics. Imputed characteristics are those that people believe to be present in things; these beliefs arise out of the innumerable messages about things that each individual becomes acquainted with through advertising and the opinions of other consumers. In the high-intensity market setting the number of messages circulating in the social environment is truly staggering. Before reaching the age of twenty a person in the United States will have been exposed on average to 350,000 television commercials.[33] In addition there are radio, billboard, newspaper, magazine, and other forms of messages. These messages constitute the symbolic or culturally determined aspects of material things themselves. Commodities are not straightforward 'objects' but are rather progressively more unstable, temporary collections of objective and imputed characteristics - that is, highly complex material-symbolic entities. The disintegration of the characteristics of objects stands in reciprocal relation to the fragmentation of needs.

NEEDS AND THE FORMS OF SATISFACTION

The tangible artifacts produced and exchanged for the satisfaction of needs represent the crystallized social form of the duality of material and symbolic correlates in human needing. This duality is present in all four patterns of social activity (the four 'pure types' described above, pp. 72-3) that provide for the satisfaction of needs. The accelerating exchange of things in the market economy - the sphere of commodity production and consumption - is based upon the increasing material complexity of the objects of satisfaction. Correspondingly, the individual necessarily becomes more and more unfamiliar with the characteristics of those objects. The duality of material and symbolic correlates is embodied in commodities just as it is in those things produced directly for use. However, as the sphere of market exchange under conditions of industrial production becomes more extensive, the duality embodied in commodities assumes an aspect of 'independence' vis-à-vis the ambiguous character of the needs that are oriented toward objects of satisfaction through any culture's socialization patterns. The 'distance' between the interpenetration of the material and symbolic aspects of needs that occurs in production activity on the one hand and in consumption activity on the other widens as these two kinds of activity become quite separate aspects of the individual's everyday experience.

Again, from my perspective the qualitative transformations that take place in the different phases of this social activity are a matter of degree, not of kind. This means that there is both a definite thread of continuity across all phases and a developmental sequence (albeit not a perfectly linear one) that shows definite qualitative turns. The duality of material and symbolic correlates in human needing is the element of continuity; the qualitative turns or points of discontinuity are related to the extent of the sphere of market exchange. The more extensive is that sphere, the greater is the degree to which that duality is embodied in material objects. In the setting of generalized exchange gradually instituted by the expanding market economy, there is an inherent tendency for individuals to develop an exclusive fixation on material objects and associated services (the complete identification of needs and commodities), at the expense of a more balanced mixture of market exchange on the one hand, and, on the other, interpersonal relations that are not mediated through market processes. An illustration of how far the prevailing tendency extends (at least in theory) is supplied by a recent study which suggests that marital arrangements and family interactions could be regarded as a form of market activity: the concepts of utility and preference ordering are used to understand the relationships among the 'services' provided or required by each of the family members.[34]

An imaginary exercise may help to clarify this perspective. Let us assume that we came upon a totally unfamiliar primitive culture and that we had no formal anthropological training. In this culture's ensemble of artifacts we would likely encounter both ceremonial objects and others readily identifiable as implements (pots, tools, etc.). The 'uses' of the first set of objects would remain a mystery to us for some time; but even with respect to the second we would be forced to learn through observation and participation the specific modes of appropriate utilization. We would have to learn in what relation they stood to each other; under what specific circumstances they could be employed; in what proportions and through what kinds of preparations or rituals the refashioned materials of nature could be enjoyed; what usages were forbidden, even though they might appear to us to represent a 'natural' or obvious application of the artifact's potential capacity for productive use; and so forth. If this culture had only limited exchanges with others, we could not fathom its general practices with respect to objects simply by accompanying its members to market. Rather we would have to be initiated into the socialization patterns and the system of interpersonal relations (the set of social 'cues') in order to understand the relations between activities and objects in that society.

Now imagine the situation in reverse. Much like ourselves in the previous illustration, a stranger transported suddenly to the industrial paradise of consumption would have to learn how to want things. To do so he would spend his days in the marketplace; but what would this mean concretely so far as his behavioural orientation was concerned? Here the symbolic mediations are extensive rather than intensive. Owing to a scarcity of time, he would have to remain relatively indifferent to any particular commodity and would have to order his preferences with respect to large classes of goods and characteristics. (Contrariwise we would have had to learn to respond intensively to the more limited range of goods in the primitive culture in order to understand the manifold layers of meaning 'sedimented' in those objects which upon superficial examination appear so simple.) Second, he would discover that the dominant social cuing in our society occurs only with reference to the ever-changing world of goods as a whole, rather than to the specific goods which are available at any particular time, and that he was expected to journey alone toward the Mecca of satisfaction, despite the fact that everyone else apparently had the same compass bearing. Were he to request the assistance of his fellow travellers, his relatively untutored senses would easily detect in their answers a varied selection from the messages inscribed on the all-too-abundant signposts along the way.[35]

Considered as a whole, the symbolic mediations that shape the expression of human needs are no more complex in industrial market societies than they are in the other patterns of need-satisfaction. There are differences, however. The mediations are extensive rather than intensive, for one thing; for another, the much greater distance between the activities of production and consumption gives a commodity a life of its own, so to speak. From the consumer's viewpoint a new commodity appears in the marketplace almost as if it were the result of spontaneous generation, only to disappear again in an equally mysterious fashion. Hegel remarked that the truth emerges from a 'bacchanalian whirl in which no member remains sober,' and the same may be said of commodities in a high-intensity market setting. Their apparently fixed character is an illusion, because each of them is only a means in the larger strategy of marketing techniques - the steady expansion of the scope of exchanges - that governs their appearance, transmutations, and disappearance before the consumer's eyes. They are in themselves only a volatile mixture of relational characteristics and possess no stable core. The characteristics that adhere today to a specific material object may not be simply transferred tomorrow to another discrete object, but may be broken up, rearranged, and distributed among a set of different objects altogether.

Throughout this section I have emphasized the thread of continuity that links the various patterns of need-satisfaction. For example, I have argued against postulating a simple dichotomy between production for use and for exchange. My objective has been to understand the high-intensity market setting not as an especially pathological phase in contrast to other phases, but as one of the many possible responses to the inherent ambiguity in human needing. The principal feature of this response is its tendency toward a complete identification of needs and commodities. And my basic intention was to discuss the problem of satisfaction not with respect to all possible responses to that ambiguity, but only to the particular one that I have been discussing.

In this setting the relationship of needs to the predominant form of satisfaction creates a state of confusion in an individual's everyday experiences. In opposition to other commentators I have argued that the individual's dilemma consists not in falling victim to false or artificial wants, but in determining the suitability of produced objects for the requirements of needs. Such a perspective carries no implication that this is the only possible social setting in which this kind of problem might arise. Even when the critical perspective is reformulated in this way, however, what entitles one to make an assertion of this nature at all? One could reply immediately that the most superficial perusal of contemporary messages about things provides an intuitive justification for it. The lowliest of things from soapflakes to shoe polish carry with them images of spontaneous delight and happiness, and presumably the very rational producers who incur considerable costs in designing and transmitting those images do not do so for entirely whimsical reasons. If we assume that those images have been found to play some significant role in modern consumer behaviour, can we say that this reveals not a pathological state of desire – for certainly there is nothing wrong with seeking happiness – but a pathological state of the objectification of desire? (By the phrase *the objectification of desire* I mean simply the widespread acceptance of grossly exaggerated claims about the importance of such great numbers of commodities for the satisfaction of needs.) If so, a friendly critic might remind us that the emphasis in this argument on the symbolic correlate of needing seems inconsistent with this judgment. For if the majority of individuals do indeed respond positively to such messages and continue to function effectively in this context, if they continue to respond positively to the general message that encourages them to orient their needs to ever-larger sets of commodities, where lie the grounds for the attribution of pathology?

To answer this objection one must find some basis for judging the appropriateness of this pattern that is consistent with the emphasis in this essay on the symbolic correlate of human needing. Otherwise one could only conclude that the high-intensity market setting is one among many of the possible routes to human need-satisfaction. In this setting the symbolic mediation of needs is different from that in other patterns of need-satisfaction. But what are the grounds for a critique of this setting?

I have argued that a critique must not rely upon a distinction between true and false needs or between production for use and for exchange. On the other hand, I have attempted to establish the thesis that, under the conditions of generalized market exchange and industrialization, many commodities become increasingly ambiguous entities, i.e., inherently unstable and temporary collections of relational characteristics. A critique can indicate on the basis of this thesis the negative aspects of the market economy in which we live.

NEGATIVE ASPECTS OF INTENSIFIED COMMODITY CIRCULATION

The high-intensity market setting has emerged as the most recent stage of the market economy that was brought into being by the capitalist mode of production. As described by writers such as Marx and Polanyi, the market economy was created by the ruthless disruption of traditional ways of treating nature and human labour in earlier modes of production. The marketplace became for the first time the basic arbiter of human destiny and of the human relationship to non-human nature.

During the earlier stages in the creation of the market economy, the working classes engaged in bitter struggles to secure the barest minimum of subsistence requirements. (This remains true today in the Third World nations whose economies are dominated by the multinational corporations.) With the growing wealth and productivity of the system attained by the early twentieth century, however, a decisive shift in the realm of consumption had to occur. Having been schooled in the production ethic of work and thrift, the majority of the population then had to be retrained as consumers. This fact was recognized by many business leaders in the 1920s, as Stuart Ewen has shown.[36] The full effect of this transition was postponed by the Great Depression and World War II, and it was not until the post-1945 era that the formation of a 'consumption ethic' became a significant feature of social policy and practices.

In the transition to this new phase of the market economy, the foundations of the capitalist mode of production remained unchanged. Ownership and control of the production process is still vested in the hands of a dominant class whose power is based on transmitting the inherited inequalities of income and wealth. Submission to hierarchical authority, exploitation, and alienated labour continue to be fundamental realities of the experience of work. What changed was only the necessity of stimulating and managing the growth of the sphere of consumption through such means as intensified advertising and the expansion of consumer credit. This new policy did not alter any of the enduring features of the sphere of production; and obscurantic concepts such as 'the consumer society' and 'consumer sovereignty,' with their implications of significant social power vested in the hands of consumers, tend to detract attention from this basic fact.

Yet the new emphasis on consumption did change important aspects of everyday life for the majority of the population. The underlying structure of production relationships persists, but there is a decisive shift in the realm of popular culture. This change has two key features: 1/ the number and complexity of available goods in the marketplace grows enormously, and the characteristics of goods change at an accelerating rate; 2/ individuals now tend to interpret feelings of well-being more and more exclusively in terms of their relative success in gaining access to higher levels of consumption. The specific objective of this essay has been not to assert that the sphere of consumption has become more crucial than the sphere of production in the capitalist market economy, but rather to define an adequate framework for understanding what happens to the articulation of needs and the sense of satisfaction and well-being under the impact of this change.[37]

To phrase the point another way, what I have tried to do is to isolate the negative aspects of intensified commodity circulation as such, thus separating this problem from associated matters such as the exploitation and alienation of labour. In one sense this is an attempt to give a more precise image of 'commodity fetishism' as it appears in the most advanced stage of the market economy. It is an examination of the problematic situation wherein the whole endeavour called the satisfaction of needs is oriented exclusively toward the realm of consumption. To the extent to which this problematic situation arises in bureaucratically administered socialist societies as well as in government-managed capitalist societies, a critique of it applies to certain features in both of these social forms.

What, then, are the negative aspects of intensified commodity circulation as such? The four most important are: 1/ a fragmentation and 'destabilization' of the categories of needing; 2/ the difficulty of 'matching' the qualities of needs with the characteristics of goods; 3/ a growing indifference to the qualities of needs or wants; and 4/ an increasing environmental risk for individuals and for society as a whole.

1/ Both in pre-modern, non-market societies and in the earlier stages of the modern market economy, there are highly structured socialization patterns that define for individuals the appropriate categories for the interpretation of their needs. Certain cultural patterns for the selection and preparation of foods, for example, generally are established for the society as a whole, and they tend to change rather slowly. What these patterns do is to shape an individual's judgments about the suitability of particular objects for particular needs - although we can assume that each individual's experience of satisfaction varies somewhat even in highly structured settings.

When the characteristics of goods change quickly and continuously, however, the categories of needing through which individuals relate to these goods are in a state of permanent fluidity. When goods become rapidly changing collections of characteristics, the individual's judgments about the suitability of particular objects for particular needs are destabilized. Characteristics are distributed and redistributed across previously distinct categories of needs, experiences, and objects. For example, the taste of menthol in a cigarette is said to be 'like' the advent of springtime and the purchase of a certain type of automobile 'like' gaining a new personality. Thus the expression of need itself is progressively fragmented into smaller and smaller bits, which are then recombined in response to market cues into patterns that are temporary, fluid, and unstable. Previous categories of need dissolve, and the resulting fragments are subjected to regular reshuffling into new patterns. The constant redivision and recombination of need-fragments renders it increasingly difficult, if not impossible, for individuals to develop a coherent set of objectives for their needs and thus to make judgments about the suitability of particular goods for them. Under these circumstances the sense of satisfaction and well-being becomes steadily more ambiguous and confused.

2/ The consumption ethic promoted today depends for its success on the destabilization of the categories of needing. In its present phase there are, however, traces of the earlier, relatively stable categories inherited from the more highly structured socialization patterns of the past. Messages about things transmitted in advertising link goods with certain images of well-being which serve as the background texture for the stimulation of desire. This background texture is often composed of traditional images of well-being drawn from social situations which have largely disappeared from everyday life: a slower pace of life, quiet and serenity, open space, and closeness to the natural environment (images of rural life); contributing to the happiness of loved ones (images of family life); attainment of goals set in accordance with personal rather than institutional demands (images of success in non-institutional settings); a sense of familiarity and security in purchasing goods (images of craft skills); a concern for quality and good judgment (images of discerning tastes).

The advertisement's composition connects background imagery with products which have not the slightest intrinsic relationship to it: the automobile or cigarette package displayed against a stunning picture of unspoiled wilderness, or the liquor bottle set in a farmhouse room full of hand-crafted furniture. Clearly the attempt is being made to transfer the positive feelings presumably evoked by the background imagery to the product which is so arbitrarily associated with it. The expansion of the industrialized market economy is in fact the main reason why those settings depicted in the background texture of advertisements have largely disappeared from the everyday life-experience of most individuals. The message plays upon the ambiguities resulting from the contradictory desires for the kinds of satisfactions associated with those vanishing settings, on the one hand, and for access to the full range of modern consumer goods on the other.

The fragmentation of needs and the rapidly changing characteristics of goods tend to undermine the more stabilized sense of well-being connected with the traditional images - incorporating images of objects and of natural environments - mentioned above. The sense of well-being arises out of a degree of success in matching the aspects of a need with the characteristics of an object (as perceived within the context of a whole social and environmental situation). This matching becomes increasingly difficult under present circumstances. Here each of an individual's many consumption choices represents a momentary resolution of a complex, unstable network of need-fragments and goods characteristics. Every provisional assortment of need-fragments is made up of bits which are also provisionally combined in many other ways; and each of the goods characteristics is also distributed among quite disparate categories of objects. If an individual feels a sense of accomplishment in a particular consumption choice, can he determine what specific bits of the need-assortment, in association with what specific characteristics of the purchased item, helped to bring this about? I think that in most circumstances he cannot, and that this also suggests why there must be greater ambiguity in the sense of satisfaction and well-being.

3/ It is not unreasonable to suppose that it is just this greater ambiguity which - at least to some extent - impels individuals to search more extensively in the array of commodities for an adequate degree of satisfaction. If this is the case, then we would have in this supposition a more accurate representation of what is usually called the insatiability of wants. And if so, then what we are witnessing is a massive auto-experiment by individuals, with the marketplace as the laboratory and their own needs and states of feeling as the experimental subject.

Given certain necessary limiting conditions referred to earlier in this essay - such as inadequate information and the scarcity of time - this quantitative expansion of wants exacts a price in relation to other dimensions in the experience of needing. The simple want for larger and larger numbers of things means that the individual must pay correspondingly less attention to the particular qualities of each want and each thing itself. In other words, the individual must become increasingly *indifferent* to the fine shadings and nuances of both wants and the objects which he pursues in the search for satisfaction. The intensive character of needing suffers in proportion to the sheer extensiveness of the search carried out among the almost infinite possibilities for matching need-fragments and goods characteristics. It seems to matter very little, for example, that in the bland uniformity of mass-produced plastic furniture we lose the subtle differences in grain and texture to be found in the wood that we now mostly turn into pulp. But this indifference to the qualities of objects, and the slightness of our attachment to them as we so readily replace them with 'new, improved' versions, is merely the other side of the developing shallowness and triviality in our articulation of our needs themselves.

4/ The possibilities for consuming so many things depend upon the hectic pace at which the accomplishments of modern scientific knowledge and technological innovation are at present being absorbed into commodity production. By these means enormous quantities of residual substances enter the biosphere. Some dangers such as lead and mercury poisoning are already widely recognized, but for the most part the existing state of scientific knowledge is inadequate for estimating the potential hazards. For example, as I mentioned earlier there is some recent speculation that the fluorocarbon propellants in aerosol sprays may have a deleterious effect on the atmospheric ozone. Whether this is the case or not, and whether the effect (if it exists) would be significant or not, cannot be established at present. What is important in a long-range perspective, however, is the potential magnitude of the unknown environmental risks that we run in using these substances, especially in such massive quantities, for the production of goods.[38] The methodical separation of commodities and actual or potential discommodities that obtains at present renders rational decision-making in this regard (both for individuals and societies) almost impossible. Again one might describe our existing practices as a massive auto-experiment of the industrialized societies, the ultimate consequences of which may jeopardize not only the biological future of the human species, but that of many other living things as well.

One would suppose that there ought to be very substantial reasons for undertaking such an experiment, but the disproportion between the potential risks on the one hand, and the shallowness of the needs in the service of which they are undertaken on the other, belies the thought. As a precautionary measure the present rate of innovation and productive application should be reduced considerably. A sober appraisal of the situation leads one to believe that even this is not a sufficient response, however. For we are dealing not with a limited and well-defined set of variables, but with the incredible complexity of the planetary ecosystems as a whole. No one can predict at what point in the development of our scientific understanding we shall have an adequate grasp of those complex mechanisms, or - given the scale on which they operate - whether we shall be in a position to devise effective compensatory measures for the unintended injuries if they are required. Prudence dictates that we reduce to the lowest possible level our measure of reliance on the potentially hazardous constituents at present being used in commodity production. The price we would have to pay for exercising this greater degree of discretion would be a sharp reduction in the number and variety of available artifacts.

These four interrelated sets of problems are the outstanding negative aspects of intensified commodity circulation that are embodied directly in the material objects themselves. The condition of endemic confusion in everyday experience which is rooted in the interplay of needs and commodities in the high-intensity market setting has so far prevented us from understanding these negative aspects and the seriousness of the potential dangers that are inherent in them.

The double ambiguity

The ambiguous character of human needing, consisting in the dual material-symbolic correlates that are interwoven by the socialization patterns which shape the interpretation of needs, is reproduced in the material objects that minister to our needs along the journey to fulfilment. In traditional, pre-industrial cultures the established socialization pattern normally imposes a general interpretation not only on the expression of needs but also on the significance of the objects, the ensemble of which tends to remain relatively stable over time. In other words, a unified set of symbolic mediations governs the interpretation of both. A widening sphere of market exchanges undermines the more immediate identification of needs and objects, giving freer play to the individual's role within a more flexible network of social constraints.

In a generalized exchange economy the market itself tends to become the chief agency through which the symbolic mediations are channelled. The market is the principal reference-framework that defines the prevailing model of rational behaviour for the members of society. Individuals in this setting have a far greater amount of freedom to experiment with their desires than is the case under conditions of limited exchange and the encrustations of long-established custom. The divorcing of production and consumption activities in a generalized exchange economy dissolves the previously unified set of symbolic mediations and produces a kind of 'brownian movement' of particular needs within the fluid medium of the market. The greater randomness of individual choices is not unstructured, for the cohesiveness of the whole is maintained through the ruling socialization pattern which encourages persons to interpret their needs solely as needs for commodities.

Industrialized production under conditions of generalized exchange makes possible a qualitative turn in the scope and flexibility of market exchanges. The capacity to synthesize new materials and energy sources in myriad ways also inaugurates a process whose outcome is the disintegration of marketed objects into transitory states of characteristic collections. The dissolution of the commodity into an unstable network of characteristics and messages heightens the degree of ambiguity in the relation between needs and the sources of satisfaction.

The fragmentation of needs stands in reciprocal relation to the dissolution of commodities as determinate objects. Both sides of this relationship have always been inherently unstable; but whereas this dual instability is normally contained within certain boundaries in 'traditional' societies due to limitations on the scope of market activities, the high-intensity market setting allows it full play. In the furious pace of daily exchanges each side tends to collapse into the other and lose its independent identity. The realm of needs becomes identical with the range of possible objects, while the nature of the object itself becomes largely a function of the psychological state of those who desire it.

Thus there arises the possibility for the extensive simulation of experience through the manipulation of messages about things. But in saying this we must keep in mind the element of continuity in the ambiguous character of human needing. The intensive cultural activity present in every phase of the sedentary life-pattern, developed and regulated through group interactions, always involves the social simulation of the experiences of individuals to some degree. The significance of the surrounding environment (topography, flora, and fauna) as well as that of created artifacts has been interpreted in every culture through established agencies, be they shamanist dances or textbooks; and to some extent in all cases - albeit with wide variations - this represents a social 'manipulation' or simulation of an individual's possible experiences. The specific difference of the high-consumption lifestyle is that it manages this process extensively (as interactions embodied in a tremendous quantity of objects) rather than intensively (as personal interactions in association with a limited range of things).

Elements of both continuities and discontinuities have been emphasized in the foregoing comparison of the high-intensity market setting with other patterns of need-satisfaction. In this discussion I have sought to identify the unique features of this setting considered as a social framework for the interpretation and satisfaction of needs. In order to pursue a single line of inquiry as carefully as possible, I have commented only in passing on the other principal aspects of this setting: the sphere of production, including health and safety factors, authoritarian or hierarchical modes of control, and the changing occupational structure of the work-force; the great domestic inequalities of income and wealth that have been institutionalized in the market economy throughout the entire course of its development; the exploitative relations between various regions of the world that arose in the periods of colonialism, imperialism, and neocolonialism.

No one of these four aspects is more important than the others. The four together form the basis of the social system in which we live. The theory of the interaction between needs and commodities that has been presented here is designed to expose the structural dynamics of one of those four aspects: the sphere of needs, consumption, and satisfaction. No solution will be proposed for the social problems that have been identified in the course of the discussion. Part Three of this essay has the more modest objective of exploring some of the dimensions in which the search for possible solutions for those problems might be undertaken.

PART THREE: PROGNOSIS

Toward an alternative setting for human needs

During the past decade many studies have been undertaken to determine whether adequate resources will be available to meet the projected material demands of the human population and especially of its most demanding part, the people of North America. The exponential rise in anticipated requirements has served as a warning to hasten the search for new sources of supply in minerals and energy. With the discovery of the need to incorporate the environmental impact of resource extraction and utilization into the overall calculations, the full dimensions of the supply problem have begun to emerge. The results for the United States have been summarized in a national government report, *Material Needs and the Environment.* According to the conclusions of this study, which have been given earlier in the preface and introduction, the 300 million people of that nation will require annually by the year 2000 many billion tons of minerals and organic materials, as well as the energy equivalent of six billion metric tons of coal.

There is every reason to believe that, given projected demands of this magnitude, anxiety about the availability of supplies will grow, especially in relation to the competing claims of the other seven billion humans who are expected to inhabit the planet by the year 2000. The terms of trade between developed and developing areas which have prevailed until now, and which began to change with the rise in oil prices, will in the future be fundamentally different where all of these much-desired substances are concerned. Increasingly contentious bickering over the prices of resources, as well as shifts in the economic and political alignments that govern effective disposition over them, will be a predominant feature of international relations.

The developing public controversy over resource availability and demographic projections has been restricted primarily to estimates of the maximum 'carrying capacity' of the planet, that is, the upper limits of the earth's ability to sustain human life at the desired levels of well-being. The proponents of the technological fix, who still command the highest respect among the ranks of government policy advisers, insist that sustained technological progress can provide a comfortable standard of living for everyone among the five, ten, or even fifteen billion humans that we may expect to occupy the planet in the twenty-first century. On the other hand, a dissenting minority claims that the limits of the earth will be reached much sooner, especially if environmental constraints - those factors involved in maintaining the viability of the planetary biosphere for human life – are carefully considered. The Ehrlichs' *Population, Resources, Environment*, published in 1970, was one of the first statements of this view to receive widespread notice; and *The Limits to Growth* (1972), a study based on highly controversial computer modelling techniques, suddenly brought this perspective to the attention of newspaper readers and television viewers. Finally, a well-known economist, Robert Heilbroner, drew the appropriately pessimistic social and political conclusions implicit in this approach in his *Inquiry into the Human Prospect* (1974).

The confrontation between these opposing viewpoints has only just begun and will undoubtedly become more heated in the future. But even now it ought to be possible to reach general agreement on three basic points.

In the first place even the most enthusiastic technological optimists must acknowledge, it seems to me, that there are some theoretical limits to economic and population growth. Further technological progress certainly can alleviate some short-term problems, but in doing so it will also only postpone the inevitable day of reckoning when controls on human economic and population expansion must be instituted. The exclusive pursuit of the technological fix is simply a way of mortgaging the future for the sake of a highly problematical present luxury, namely avoiding the acceptance of responsibility for the potentially disastrous future consequences of policies adopted now. The pompous bluster of the technocratic futurologists is in part a mask for political cowardice and impotence.

Second, as far as resource demand and possible environmental degradation are concerned, aggregate human population growth is *not* the crucial element of our dilemma. The only adequate calculation for the problem of human population growth as such is the one that is derived from the following equation: the environmental impact of human demands equals population times consumption of goods per person times the environmental impact per quantity of goods consumed.[1] The one-third of the human population in the industrially developed nations currently uses ninety per cent of the available resources; it is the exponential increase and environmental impact of *their* demands, not those of the human population as a whole, which is the real and immediate cause of the emerging global crisis.[2] In saying this I do not mean to deny that aggregate population expansion itself represents a dilemma for human society. One could argue from a variety cf perspectives that existing populations everywhere on the planet - in developed as well as developing nations - are in excess of desirable levels. This is not the immediate problem, however. It is the unforgivable squandering of resources in the developed countries and the environmental degradation resulting therefrom that currently determine the general direction of the global 'political economy' and that constitute the source of potential future disasters for the entire human population. Whether we admire the siren song of complacent technological optimism or not, we should at least recognize who composed the tune.

Third, the massive scale of the forces now at play has created an inertial factor that constrains decision-making possibilities. In other words, so great are the dimensions of the existing practices that their impact would continue to be felt far into the future even if dramatic changes in the direction of policy were introduced now. For example, centralization of population and production is ensuring a future dependency on high-level energy use. The concentration of the propellant gases from aerosol containers in the upper atmosphere would not peak until 2000 even if a total and immediate ban on its industrial and consumer applications were implemented at once. And were we to achieve now a stabilization of human reproduction at the replacement level (net reproductive rate of 1.0), the aggregate population size would continue to grow for over fifty years. Since these are wholly unrealistic presumptions, the actual results which policy changes could hope to achieve would be far more modest. The inertia of the immense institutional and environmental forces, together with the complexity of the interrelationships among them, raises the possibility that in the future we may be faced with more and more problems that are past the point of easy solution by the time they are recognized.

The inertia and complexity of the forces at play in the global political economy should convince us that it is unwise to wait until we are pressed against the limits of the planet's carrying capacity before we seek to bring under rational control the social dynamic that is leading us inexorably thither. Since it is the political economy of the developed nations that is the motor of this social dynamic, and since it is a model of 'development' originating therein that is extolled to the rest as a universal ideal, we must examine the social practice based on that model if we wish to assess the possibilities for redirecting its course. That social practice has evolved most completely in North America, which among the ranks of developed countries has marked out the path for the others, including some of those that are nominally socialist. Its hallmark is the ideology of insatiable want and unlimited economic growth.

In this essay I have argued that the debate over the boundaries of the planet's potential carrying capacity encompasses only one aspect of the increasingly grave social dilemma that confronts us. The restriction of debate to this single aspect of what is a multidimensional problem is dangerous because no definitive proof concerning those boundaries will be forthcoming until it is too late to reverse the direction of events (on account of the 'overshooting' effect) without experiencing widespread social catastrophes. We must expand the parameters of the debate. We must confront explicitly the ideology of insatiable want and unlimited growth and the social practice that sustains it.

The actual form of this social practice is what I have called the high-intensity market setting, which leads to a situation in which individuals are encouraged to interpret their needs exclusively as needs for commodities. Among the range of developed nations this situation takes diverse forms, including capitalist and socialist variants, and exists in quite different stages of evolution. The behaviour of the affluent minority in the developed nations is regarded in this essay as an 'ideal type' of the high-consumption lifestyle. Obviously many persons in the developed countries are debarred from participation in this lifestyle. However, I have assumed that this ideal receives a significant measure of support (active or passive) from the majority of the population and therefore that it is a crucial factor in shaping the general trend of social change. In *this* context the fact that the material benefits of the system are distributed so inequitably is unimportant.

This ideal encourages all individuals to place consumption activity at the very centre of their everyday concerns, while simultaneously reinforcing the experience of dissatisfaction at every attained level of consumption. In this vicious circle of expanding gratifications and frustrations all the technocratic fantasies of untold future wealth are simply irrelevant to the problem at hand. It is the social practice of the high-intensity market order, institutionalized to varying degrees in the developed nations and promoted elsewhere (by economic coercion and ideological warfare) as a universal ideal, that is responsible for having dissipated much of the planet's readily available resources, for making the survival of the existing population dependent upon energy-intensive agriculture, for creating the necessity of future reliance upon nuclear energy (with its attendant dangers), and for introducing massive quantities of toxic compounds into the biosphere.

There is every reason to expect that in the immediate future we shall continue to follow this path and to test in theory and practice the limits of the planet's carrying capacity. But we cannot postpone indefinitely the opening of serious public debate on the inadequacies of the social practice that causes us to tempt fate in our ecological situation, especially in view of the inertial force of our present policies. And any hope of altering our course depends in the first place on achieving a clear understanding of our present heading.

This essay has attempted to provide some guidelines for that process of self-clarification, and a basic methodological principle has been employed in it. In trying to expose the interior structure of a dynamic process - the interplay of needs and commodities in a specific social framework - I have deliberately avoided any attempt to judge that process according to some positive, abstract standard. This approach has been taken by those who formulate a scheme of 'basic human needs' and then seek to measure the practices of a given society against those established criteria. I have noted earlier some of the deficiencies in what one might call this 'positive theory of needs.' There is a great difficulty in organizing any scheme that will receive general assent; in any case the needs thus specified tend to be so broadly conceived that a great variety of behaviour patterns can be subsumed under them. Most theorists concede in any case that the actual definition and elaboration of needs can only be given by the individual and collective activities of the members of a society.[3]

A critical perspective can, however, uncover the possible contradictions in a system of needs, i.e., the layers of confusion which prevent individuals from realizing their own objectives and potentialities. This kind of approach, which has been employed in the foregoing discussion, is characterized by a 'negative theory of needs.' It does not attempt to specify what are or should be the genuine or appropriate desires of individuals in the particular historical circumstances of the moment. Its basic objective is to show why the expression of needs becomes increasingly incoherent in a high-intensity market setting and how individuals become unable to understand the objectives of their wants. This approach most certainly does not presume that any other form of social practice (past or present) provides a model for the expression and satisfaction of human needs that is superior in every respect and is applicable to every situation. It does attempt to identify the specific form of the set of contradictions in a specific social practice. The theory of needs which informs the analysis is 'negative' in the sense that it refrains from defining the categories that might be appropriate to a coherent network of needs. Only the individuals and groups who discover for themselves the inadequacy of the existing system can provide those categories.

This negative theory of needs is not entirely agnostic, however. For while one must not presume to define the needs which would emerge in an alternative setting, one can postulate the necessary conditions for the emergence of alternatives. There are two conditions: the institutional reorganizations required to bring about a conserver society; the practical understanding of human needs in relation to the needs of other living entities in the biosphere - indeed, in relation to non-human nature as a whole.

No elaborate argument should be necessary to establish the point that there are some limits to economic and population growth. But everything depends upon whether we regard such limits as a bitter disappointment or as a welcome opportunity to turn from quantitative accumulation to qualitative improvement in the course of creating a conserver society. If it is to be regarded as a welcome opportunity, then it is not an end in itself, but only a necessary framework for certain alternative practices which can provide a new setting for the expression of needs. Some traditions of social thought during the nineteenth and twentieth centuries have suggested ways of redirecting social change which can be evaluated in this context.

At what level of production and consumption should an alternative setting be instituted? In the first place, this question cannot be answered abstractly; and in the second, I think that we can give a more adequate answer to it if we do not remain within an exclusively anthropocentric framework for the understanding of needs. I suggest that we should explore the possibilities for building a syncretic framework in which human needs are conceived in relation to the needs of non-human nature. This is not a plea for a new mysticism. In fact such a framework is not only consistent with our scientific understanding of nature, but is a logical extension of that understanding. It is, as I will try to show, the framework that is most appropriate for the values which would govern the transition to a new setting for the expression of human needs.

The operative assumption in the following remarks is that the possibilities for progressive social change in the developed nations are dependent upon breaking the upward spiral of needs and commodities. In my view this is a necessary condition for progressive social change in these nations, but quite obviously it is not a sufficient condition of itself. Transcending the phase of the high-intensity market order is impossible to conceive without also contemplating other necessary changes. At least two other changes are essential, namely a reduction in the economic and political power of multinational corporations and a more peaceful state of international relations. The existing system did not appear spontaneously, as it were, but rather evolved as a function of concentrated economic power; thus it is unthinkable that the former could be overcome without confronting and overcoming the latter. Similarly, the developed nations have gradually become tightly interlocked in their economic relations (recently this has also included the relations between capitalist and industrialized socialist countries), with the result that changes in one nation that are opposed by some of its economic partners might be sabotaged by manipulations in currency exchange and capital flows in the first instance, and perhaps by political and military measures if those others fail.

Thus all speculations about ways to overcome the destructive features of the existing socio-economic order remain 'utopian,' for the obvious reason that no practical program can be given for undermining the power of the existing international economic and political arrangements which sustain that order. If this is indeed the case, what can possibly justify any such speculations? Why should we entertain the prospects for bringing utopia down to earth by discussing alternative policies that would require a substantial modification of our present practices?

The basis for doing so is the reasonableness of an expectation that the social policies upon which the prevailing political wisdom is founded may be declared bankrupt sometime within the next quarter-century. No argument can be offered to prove that this expectation is well-founded. It is certainly possible that the developed nations may take rigorous and successful measures to maintain their disproportionate share of the earth's resources; that (if need be) the affluent minorities in these countries may devise authoritarian domestic policies to retain their control over the direction of social change; and that environmental degradation may be contained within 'acceptable' parameters within the territories of these countries.

But it is not *prima facie* absurd to assume that even the most strenuous efforts may be incapable of containing the course of events in this manner. Heilbroner's *Inquiry* has offered some reasons for accepting a more pessimistic forecast, and even a fictional prognosis such as

John Brunner's *The Sheep Look Up* appears less and less unrealistic in comparison with the cheerful technological fantasies of those like Herman Kahn, whose buoyant messages are still received with awe and respect in the councils of government. I have suggested that the choice of future policy options is not entirely a matter of one's subjective temperament, however. For it is not reasonable to believe that there are no limits to economic growth, population expansion, and resource-energy utilization. Then why should we believe that the most sensible and realistic policy is to continue along our present course until we are forcibly debarred from doing so? There is very little evidence in previous human history to indicate that established institutions can respond adequately to the sudden appearance of new constraints or limiting conditions; in fact, just the opposite appears to be the case. Present-day institutions are far more massive than most earlier ones and consequently have a far greater inertial force, and a monumental effort will be required to redirect them.

Addressing himself in 1848 to future generations, John Stuart Mill wrote: 'I sincerely hope, for the sake of posterity, that they will be content to be stationary, long before necessity compels them to it.'[4] Unfortunately the moment of necessity has arrived well in advance even of any serious consideration of the question about whether we can be content under such circumstances.

The conserver society

John Stuart Mill included a brief chapter entitled 'Of the Stationary State' in his *Principles of Political Economy*, which advocated a stabilization of economic progress and population growth. In it he proposed the audacious notion that the productive capacities and population level existing at that time need not expand any further. Of course he recognized the gross inequities in the distribution of life's amenities and the opportunities for self-fulfilment among the population, but he contended that these could only be remedied by more rational forms of social organization. He maintained - correctly, as it turned out - that quantitative increases would not necessarily improve the lot of mankind as a whole. Finally, he argued that a turn from quantitative to qualitative criteria was the primary desideratum for further social progress.

For over a century thereafter the predominant traditions of social thought have maintained a conspiracy of silence with regard to Mill's suggestion. Recently it has been resurrected and applied in the context of contemporary issues.[5] This recent discussion, however, only tends to confirm Mill's fear that the idea of the stationary state would be taken seriously only in response to pressing necessity, rather than being freely adopted as a desirable framework for qualitative improvement. In a situation of bitter competition over the disposition of resources and wealth, the coming of the stationary state may occasion heightened conflict and attempts at authoritarian political solutions, as Heilbroner has intimated.

The other alternative - the expectation that the stationary state would afford an opportunity for qualitative improvement - remains hidden in the subterranean regions of contemporary discussion. Mill himself did not indicate what kinds of innovative social practices might be appropriate for the positive accomplishments of the stationary state - or, to use a more adequate current expression, for the objectives of a conserver society.[6] Suggestions that are relevant to this point can be found in the writings of other nineteenth- and twentieth-century thinkers, even though these suggestions were not advanced as a direct response to Mill's idea. They may be found in the works of earlier thinkers such as Fourier, Marx, Ruskin, Morris, and Kropotkin (as well as in other writings by Mill), and in the works of contemporary authors such as Bookchin, Fromm, Illich, Goodman, Macpherson, and Marcuse. I wish to refer to them not to advocate the adoption of any specific practical proposals, but only to illustrate some general characteristics of one set of possible alternative directions for social change.

The positive ideals and values that have informed this alternative tradition of thought have always centred on the emerging dimension of human individuality. These thinkers were never exclusively preoccupied - as I have been in this essay - with the negative aspects of the high-intensity market setting as such. The common focus of their positive ideals has been the assertion that a social transformation could make possible a rich dimension of genuine satisfaction in both the labour activities and the free time of all individuals. If there is one basic proposition that forms the cornerstone of this tradition, it is the idea that the possibilities of human satisfaction must be rooted in the creation of a well-functioning sphere of shared activity and decision-making within which individuals would forge the means for satisfying their needs.

For this tradition qualitative differences in the mode of organization of socially necessary labour, and the relation of this labour to play and leisure activities, are the decisive element in the problem of satisfaction that arises in various forms of production and consumption activities. The organization of labour through a network of non-hierarchical, community-based associations, for example, would constitute a concrete form of social practice very different from the prevailing one. Such a decision-making structure, it is argued, would allow individuals to determine their needs under conditions of freedom and autonomy. In other words, what is most important for this tradition is a change in the mode of expressing and satisfying needs, not a definition or predetermination of an alternative set of needs as such. As the outcome of qualitatively different conditions of life, the new structures of needing could not be specified in advance. The 'negative theory of needs' mentioned earlier is thus an essential element in this tradition.

In this perspective the possibility of satisfaction would be primarily a function of the organization of productive activity, and not - as in our society today - primarily a function of consumption. There is some evidence to show that individuals are aware of the fact that increasing rates of consumption are supposed to compensate for frustrations experienced in other domains of social life.[7] The argument of this essay, which has emphasized the ambiguities inherent in the consumption process, has tried to show why this is a fruitless endeavour. If this argument is correct, then we have reason to believe that the possibilities for progressive social change are dependent upon identifying other spheres of activity, apart from the consumption process, in which the sense of satisfaction and well-being also may be grounded. A full-length study would be required in order to provide a critical review of the proposals made in this regard within the tradition of thought under discussion; this cannot be attempted here.[8] A few brief remarks must suffice just to indicate the tenor of these proposals.

On the level of individual activity, the possibilities of satisfaction in an alternative setting would be related to overcoming the extreme specialization of work in the provisioning of goods and services (especially the methodical separation of mental and manual labour) and a consequent limitation of the sphere of commodity exchange. This has been a consistent theme in the tradition, and recently Ivan Illich has given it a new designation, *conviviality*. Illich uses this term to describe a social fabric based upon 'autonomous and creative intercourse among persons, and the intercourse of persons with their environment.' The goal of a convivial society is gradually to dismember the massive institutional structures of the industrialized economy and to reduce, so far as possible, the dependence of individuals on them: 'People have a native capacity for healing, consoling, moving, learning, building their houses, and burying their dead. Each of these capacities meets a need. The means for the satisfaction of these needs are abundant so long as they depend primarily on what people can do for themselves, with only marginal dependence on commodities. These activities have use-value without having been given exchange-value. Their exercise at the service of man is not considered labor. These basic satisfactions become scarce when the social environment is transformed in such a manner that basic needs can no longer be met by abundant competence.'[9] The current organization of producing and consuming activity inhibits the development of the individual's talents and capacities for participating directly in activities that provide the means for satisfying a wide range of needs (building a home, growing food, making clothing), and instead orients his activity exclusively around market purchases. The convivial society would encourage individuals to engage directly in productive activities to the greatest degree possible.

There is nothing inherently evil in commodities and market exchanges as such, and there is no reason to believe that it would be desirable to extirpate them completely. There is cause for concern only when commodity exchange tends to become the *exclusive* mode for the satisfaction of needs. Nor is there any profit in attempting to specify a universally valid pattern for the relative significance of the sphere of market exchange in the overall economic calculus of every society. We may assume that the degree of significance of commodity production will vary widely according to specific historical circumstances and variations in social organization and personal desires. There need not be any uniform pattern even within a particular society, if that society is sufficiently decentralized to allow wide variations in the choices available to its members. The simple rule governing such choices may be stated as follows: reliance on access to complex manufactured objects through market exchanges is inversely related to the degree of local, direct control over the means of satisfying needs.

Obviously these choices of varying patterns could only become realistic alternatives after the present national concentrations of wealth and wide regional disparities had been overcome as a result of deliberate social policies. Promotion of a stultifying uniformity in types of material objects and life-situations, both within and among nations, is one of the primary tendencies of generalized market exchange in industrialized economies. This is not a necessary result of technological progress itself, but is rather a function of the social policies that favour massive, centralized technologies of production over intermediate and small-scale technologies. The pattern of public and private investment which results from such policies concentrates productive resources and undermines the viability of any social and economic activity outside the metropolis that is not directly connected to the organizing centres of production.

The objective of an alternative social policy would not be to return a larger portion of the population to the harshness of circumstances which in the past often characterized life in the hinterland, but to disperse the advantages of modern technology - deliberately sacrificing some of the dubious 'efficiency' of centralized production - over a wider variety of situations.[10] This exercise would indeed be self-defeating if another uniform pattern - for example, identical smaller-scale communities - were to be substituted for the prevailing one. At present life in a rural community is often simply a poorer version of the urban style. In order for a variety of equally viable life-situations to emerge, ranging from the city to the small community, a significant proportion of the investments currently favouring the concentration of production and generalized commodity exchange would have to be diverted to the support of alternative patterns of social organization.[11]

The 'ideal type' for existing social policy is a population which is concentrated in urban high-rise dwellings and is dependent upon massive, complex systems for its energy supply, food and other necessities, and waste disposal. Every aspect of this life situation is energy-intensive: heating and cooling of buildings, mechanized agriculture, transportation, manufacturing of commodities, and recycling of waste products.[12] Huge investments for the continued provisioning of fossil fuels and now for nuclear generating plants will be required to sustain it. Those investments could be redirected to the support of efficient technologies for alternative ideals, for example the small dwelling or group of buildings which relied upon solar and wind energy sources, greenhouses, and the recycling of nutrients from wastes to provide a high proportion of the basic requirements of daily life. Further adaptations of new technologies in communications and local computer terminals linked to information storage banks could overcome some of the disadvantages formerly imposed upon scattered smaller communities.

The objective of an alternative social policy would not be to force all individuals to adopt a different uniform pattern of living, but rather to make other options more attractive than they are at present. Now investment decisions have the effect of making a single option – the high-intensity market setting of the concentrated urban population – the only viable one. But under alternative policies the positive features of industrialization and sophisticated technologies could offer to contemporary society a luxury not available to earlier societies, namely the ability to sustain a variety of life-situations that are more equally attractive for individuals. Different patterns of needing and satisfactions, responsive to differences in the needs of individuals, are a real possibility under present conditions. If the genius of modern productive technology were applied to the task of dispersing material and cultural accomplishments more equitably across a variety of life-situations instead of forcing them into a single mould, a wider range of realistic choices would be available to individuals. Under these circumstances individuals could choose, for example, to trade off at diverse levels the satisfactions gained by personal activity in the production of everyday requirements against the satisfactions of consumption in a generalized market setting.

The directions established by general social policies will fix the parameters within which the trade-offs made by individuals will occur. In other words, the relative attractiveness of different options for individuals will be determined in part by the options available to the society as a whole. The postwar rise in the standard of living in the industrialized nations was to some extent based on the availability of cheap energy and minerals; that situation is fast changing. The rising costs of energy, material resources, and environmental degradation soon will begin to constrict the promised delights of the high-consumption lifestyle. To sustain the mirage of ever more extensive forms of satisfaction there will be a progressive deterioration in the quality of goods and – following Martin Krieger's scenario – an expansion of simulated gratification. In this context alternatives to the forms of satisfaction that are characteristic of the high-intensity market setting may appear more attractive, so long as they are not repressed by a social system which remains intent on enforcing the hegemony of that one path to happiness, no matter what the cost may be.

Opponents of the alternative tradition under discussion here have often castigated its ideals and values as 'primitive,' 'romantic,' and 'reactionary.' Therefore it is essential to stipulate at the outset that this perspective does not seek to glorify any earlier state of social development as a golden age to which we should return. It does not necessarily assume that any other earlier social pattern achieved a better realization of that 'autonomous and creative intercourse' among persons and between persons and their environment of which it speaks. What it has suggested is that the present form of producing and consuming activity (including our dependence on energy-intensive, large-scale industrial technology) discourages the growth of personal autonomy, creativity, and responsibility. To be sure, some scope for the expression of these traits has existed in most forms of social organization, including the one analyzed in this essay; moreover, some of the achievements of modern industrialism have opened up new possibilities for their expression and realization. Thus we need not adopt the stance that our existing society represents a scene of unrelieved repression of individual self-fulfilment nor that any set of alternative proposals promises an instant and complete cure for all the ills of humanity. My own assumptions here are simply that the time when we could afford to indulge ourselves in certain illusions about the prevailing pattern is fast running out, and that recalling the challenge of an alternative tradition can broaden the range of the future options which we may wish to consider.

At present the direction of social change is governed by the ideal of an unlimited expansion of material production and its supporting infrastructure (more elaborate large-scale technologies, higher energy demands, centralization of production and population, increasing specialization of functions, and a wider assortment of commodities). I have argued that the negative aspects of this ideal introduce certain dangers whose potential dimensions are so vast that it may be impossible to deal with them effectively once their nature becomes evident. These negative features of the dominant ideal are inherent in its very structure and are magnified in direct proportion to the success and prosperity of the high-intensity market setting. Should that system begin to falter as a result of certain constraints, such as the rising costs mentioned earlier, other difficulties would arise.

In general there are two possible ways of responding to this situation. The social ideal of the high-consumption lifestyle may be upheld in the face of emerging constraints for as long as possible. Existing inequalities would persist and perhaps worsen as the overall standard of living - measured in terms of the accumulation of commodities -

ceased to rise or began to decline. Since the operative principle of this social ideal is the tendency to channel all desires and aspirations toward the sphere of consumption, anxieties and frustrations would be exacerbated as the realm of promised satisfactions shrank. Social tensions would increase rapidly, because the general level of material expectations is so much higher now than it was in the past, for example in the period preceding the Great Depression. To maintain that ideal under such circumstances would inevitably require the imposition of more repressive and authoritarian forms of political rule, managed by the agents of the privileged minority.

Another way is open, although greater courage and foresight are necessary in order to follow it. The talents and instrumentalities required for undertaking the journey already exist. As indicated above, one aspect of this way involves the development of alternative situations in which more direct and personal participation in activities related to the satisfaction of needs is encouraged, in opposition to the prevailing tendency to route all aspects of needing through the increasingly ambiguous world of commodities. If this alternative is to have any chance of success, the first steps in this direction should be taken now, before the accumulating confusions and frustrations that arise in the existing system produce a mood of sullen despair among the citizenry.

The specific forms of such activities cannot be displayed for selection in the pages of a theoretical treatise, as if what we required were a different version of the department store catalogues that appear regularly in our mailboxes. The intention of the critical analysis undertaken in this essay is to detail the structural faults of the social practice that orients needs exclusively toward commodities; a redirection of social policies and capital investment so as to reduce our reliance on that mechanism is essential. The result would be an attempt to satisfy our needs in ways that require far less energy and fewer material resources. The new routes would vary considerably in relation to socio-economic, political, technological, cultural, and geographical peculiarities. They cannot be mapped in advance as an inducement to the undertaking. They would be quite different from any of the detailed prescriptions found in the alternative traditions of thought, such as the one that has been mentioned for illustrative purposes in the foregoing remarks. But I cannot believe that we lack the ingenuity to construct any such routes at all.

After experiencing another century of 'development' we can give a more precise general guideline for a decisive shift in the direction of social change than could Mill. The conserver society is the ensemble of social policies in the industrially developed nations whose objective would be to lessen the importance of commodities as factors in the satisfaction of human needs and to minimize per capita energy and materials demands. Most future technological innovation would be designed to assist the realization of this objective and to combat the accumulated effects of residual industrial wastes in the environment. One further point is especially important: the transition to the conserver society would represent a form of social progress only if a series of related policies, such as the gradual overcoming of the endemic poverty that persists even in the richest industrialized nations, were an integral part of the objective stated above. If this were not the case, quite obviously the conserver society would simply be another form of impoverished existence for the disadvantaged elements in society.

The two basic points I wish to make about the idea of the conserver society may be summarized as follows. First, it is not an end in itself, but rather an organizing frame of reference for a reorientation of social policy away from quantitative criteria of well-being and toward qualitative criteria. Therefore it should be conceived positively as an intensely dynamic phase of social reorganization, rather than being viewed negatively as the sign of defeat for our cherished ambitions. Second, it is unimportant whether at any particular time the conserver society is characterized by an economic situation of growth, stabilization, or decline in the Gross National Product, for this situation must vary with particular circumstances and needs. The primary desideratum is a reallocation of resources and a reorientation of policies so that the problem of the satisfaction of needs is no longer viewed exclusively as a function of consumption activity. There is a growing tendency among individuals and social agencies, for example, to pay more attention to job satisfaction and shared decision-making in production activity.[13] This is the kind of tendency we should strive to encourage.

Mill stipulated that a decent measure of social justice must be a feature of his stationary state. Not absolute social equality, but a 'well-paid and affluent body of labourers,' a limitation on the accumulation of wealth, and a modest income for those engaged in intellectual pursuits constituted for him some of the social policies to be implemented. Population would also be stabilized: Mill thought that already in his day the desirable level had been reached almost everywhere on the planet. But there is another dimension as well: 'Nor is there much satisfaction in contemplating the world with nothing left to the spontaneous activity of nature; with every rood of land brought into cultivation, which is capable of growing food for human beings; every flowery waste or natural pasture ploughed up, all quadrupeds or birds which are not domesticated for man's use exterminated as his rivals for food, every hedgerow or superfluous tree rooted out, and scarcely a place left where a wild shrub or flower could grow without being eradicated as a weed in the name of improved agriculture.'[14] The basis of Mill's plea was an aesthetic notion, shared by a number of thinkers in his day, which expressed their belief that the finer sensibilities of mankind were tutored by the experience of undomesticated nature. Today there are different reasons for recalling Mill's concern for the relation of man and nature in the context of a conserver society. The relentless expansion of industrial production and human numbers has rudely shifted the locus of concern from aesthetic education to biological survival. The problem of the impact of human demands on the natural environment has now reached such a magnitude that we must consider the question of human needs as an integral part of a larger network of ecological interactions.

The needs of non-human nature

In the section on needs I noted that the discussions of human needs in psychological, economic, and sociological literature have remained within a narrowly anthropocentric framework. In other words, these discussions consider the many problems raised by the quantitative and qualitative dimensions of human needing only in relation to goals that are thought to be appropriate for a decent state of human existence, without also considering the impact of the activities undertaken in pursuit of those goals on the viability of other life-forms, and on some of the common requirements of other living things in the biosphere upon which we depend for our existence. Some anthropological contributions have sought to understand patterns of human ecology in relation to wider environmental interactions, but these are exceptional cases. I believe that we must now abandon that narrow anthropocentric framework and develop a comprehensive approach that searches for specific ways of adjusting human demands to the general requirements of other life-forms in the biosphere.[15]

In this context I mentioned the concept of homeostasis, which describes some of the processes governing the relationship between a species and its environment. These processes involve a complex set of feedback loops through which the population of each species is continually adjusted in response to competing populations, seasonal and long-range climatic variations, available food resources, and occasional local catastrophes. In a natural climax community both the number and diversity of species provides a complex feedback system which will tend to damp any perturbations.

If we view the evolutionary course of the human population in this context, we may understand it as a series of successful adaptations that permitted this species to repress the operations of the usual feedback loops. The first may have been the efficient social hunting techniques that contributed to the extinction of its prey in the Pleistocene period and that marked man as a uniquely proficient predator. Later techniques of plant cultivation and animal domestication increased the population density, and the collective power derived from permanent large-scale settlements tipped the balance further in man's favour. Finally, a new form of social organization, in which modern science and technology matured as the instrumental foundation for the so-called conquest of nature, introduced a further qualitative change in the relation of the human population to the environment. In this general pattern there were of course many regional fluctuations and intermediate stages.

None of this means that all environmental feedback loops were eliminated so far as the human population was concerned. Rather it is the common feedback pattern, constituting the homeostatic process for many species, which was repressed or suspended, as it were. A unique human pattern, governed increasingly by technological manipulations, emerged in each of these stages. In that unique pattern there are both positive and negative features.

The positive aspect is a rich elaboration of human culture as a whole, including the sphere of needing. This is made possible (at least in part) by the modifications of the normal patterns in environmental feedback loops. In human development there is a struggle against the immediate 'givenness' of environmental circumstances, in other words against interpreting the realm of satisfaction narrowly. Human societies interweave an intricate set of mediations - collective and culturally transmitted acts of reflection and interpretation - with a broad range of options in the possible sources of gratification. In the subtle transformations within the experience of needing there is an inherent resistance against a purely passive adjustment to environmental opportunities and constraints; and out of that resistance develops the human need to determine its interaction with the environment according to its reflective understanding of a holistic 'system of needs.' In other words, the symbolic mediations transform a series of discrete individual or group requirements into a cultural totality extending intergenerationally across past, present, and future.

Each human society's response to its environment is conditioned by that totality, rather than by a direct reaction to immediate circumstances, and it strives to preserve the integrity of the cultural totality against fluctuations in environmental conditions. The positive aspect of this process is the richness in the sphere of needing that is a universal feature of human culture - for example, the beautiful rendering of common artifacts present even at the lowest levels of material satisfaction. This richness is primarily intensive rather than extensive. It is not a simple function of the aggregate quantity of material objects, but rather is expressed in the complex levels of significance within which the articulation of needs and the understanding of man's relation to the rest of nature occur.

The negative aspect of this process is revealed in conceptual terms in the affirmation of the 'separateness' of man from the rest of nature, in an attitude of arrogance with regard to the manipulation of his surroundings. The ideology of human domination over nature is the most extreme version of this attitude. Its practical expression is the sort of pursuit of human goals in production and consumption activities that remains utterly indifferent to the impact of those activities on the viability of other life-forms; the most extreme version is the social practice based on the doctrine of the insatiability of human material needs. In this social practice non-human nature is regarded as nothing but a means for human satisfaction.

The result of this practice, I argued earlier, is that persons become increasingly indifferent to the character of their own needs and become entangled in a web of ambiguity so far as the relation of their needs to the objects that are supposed to service them is concerned. At the higher levels of economic development based on this orientation - that is, in the high-intensity market setting - the industrialized, energy-intensive production of commodities threatens to bring about environmental degradation of incalculable dimensions. The social and political order becomes 'hooked' on perpetual economic growth and on the necessity to take the increasingly grave environmental risks that stem from the complex production process which sustains growth.

The chief negative aspect of the process under discussion - the modification of the normal environmental feedback loops - is the difficulty in bringing about a practical realization of the need for the rational management of human demands in relation to the tolerance limits of the biosphere. Let us restate the assumptions outlined above for this discussion: there are *some* tolerance limits in the biosphere; the industrial production systems of the developed nations are now testing those limits; we do not know what these limits are at present; it is unwise to continue along our present path until we reach or overshoot these limits, since by that time it may be impossible to mitigate the adverse effects or to do so only at the cost of catastrophic social disruptions.

Simply stated, my point is that it is now a matter of enlightened self-interest for us to reassess the impact of our material demands on the environment. This is primarily the responsibility of the industrially developed world, since its practices are the cause of the incipient crisis, the possible effects of which, for example the pollution of the atmosphere and the oceans, threaten the well-being of all peoples. Our existing scientific knowledge enables us to recognize that there are some objective requirements for the maintenance of the biospheric processes upon which we depend. I have referred to these requirements in a metaphorical way as 'the needs of non-human nature' in order to assimilate this larger dimension into the calculus of human needs. For our most urgent need at present is to attempt to comprehend those requirements and to adjust our material demands in accordance with them.

Enlightened self-interest dictates that we reallocate the priorities of our scientific and technological research capabilities. There is great urgency in the investigation of the possible impact of aerosol propellants on the atmospheric ozone, for example, but the resources now committed to it are a mere fraction of those still being applied to the pursuits of the marketplace, which undoubtedly are preparing similar dilemmas for us. Vinyl chloride, asbestos, lead, mercury - some of the dangers of these widely used substances are already known, but there may be many more to be discovered, together with others as yet unsuspected. Enlightened self-interest requires us to consider most deliberately the wisdom of a general dependence on nuclear energy, which will indeed become necessary to sustain the high-consumption lifestyle of concentrated urban populations by the end of this century. The immensity of the biological risks should ensure that we give a fair hearing to alternative options, for example the encouragement of lower-density communities utilizing solar and wind energy sources to the greatest possible extent.

A program of enlightened self-interest can be founded in our developing scientific understanding of the general requirements for maintaining the viability of human life on this planet. The first steps in this direction have been undertaken in the past few years. An illustration is provided by a recent scientific conference on 'biotic impoverishment,' i.e., the accelerated rate of disappearance of plant and animal species under the pressure of recent human economic and population growth. The report on the conference cited the following problem: 'If formations in the Amazon continue to be cut over, for example, it is estimated that some one million species of flora and fauna will disappear (the worldwide extinction rate up to now is estimated at about 10,000 species per century).'[16] Species become extinct in the normal course of evolution, irrespective of human intervention in the environment; what concerns some scientists is the possible adverse consequences (indeterminable at present) of the dramatic leap in the rate of extinction that is directly related to the human pressure on the natural environment. Their assumption is that a certain degree of biotic diversity on the planet is necessary for the continued viability of the complex forms of organic life. At the conference there was some discussion of the need for an 'ethic' of biotic diversity in which 'such diversity is perceived as a value in itself and is tied in with the survival and fitness of the human race.' The idea that we might adopt this ethic for its intrinsic value is perhaps too sanguine, but at least prudence and enlightened self-interest counsel us to investigate its relevance for our own biological future.

An ethic of biotic diversity is an example of a principle that would relate human need to the needs of non-human nature. In other words, such a principle would locate human need within a wider reference-framework, beyond a shortsighted anthropocentric perspective, on the basis of enlightened self-interest. Pre-modern societies also had reference-frameworks in which forces beyond human control were recognized as operating in nature, and in accordance with these conceptions men and women were prompted to check their perceived interests against a wider set of interests or purposes. These frameworks provided a total world-view within which the appropriateness of human actions was judged. However, if any such wider reference-framework were to emerge in modern society, it would have to be consistent with the understanding of nature that lies at the basis of modern science. This does not mean that science holds exclusive rights with regard to the interpretation of nature; but the program of enlightened self-interest, as a practical matter, is dependent upon the further research of that prodigious science and technology with whose assistance modern society has caused us to court biological disaster in the first place.

From the viewpoint of our enlightened self-interest, therefore, the idea of the 'needs of non-human nature' refers only to the functional parameters within which the common interests of many life-forms are defined. Presumably, the condition of the atmospheric ozone is one example of those basic parameters. Our scientific understanding of nature can attempt to chart these parameters; and where it is unable to do so with any precision, it may publicize the inadequacy of our knowledge and counsel us to proceed with extreme caution in our impact on the chemistry of the biosphere.[17] For despite our potent technological ingenuity, we may find that we are unable to mitigate the adverse consequences of certain actions before they have exacted a terrible price with regard to the continued viability of human and non-human life-forms.

Enlightened self-interest is simply a matter of taking a long-range rather than a short-range view of human prospects, using both the attained knowledge and the precautionary counsel of our scientific approach to nature. The attempt to understand the complex processes of major ecosystems is difficult enough, but equally problematic is the task of devising effective institutional means for translating this scientific knowledge (and the awareness of its existing inadequacies) into appropriate social policies. In specific terms, how can we make an ethic of biotic diversity operative, for example? Modern secular institutions, both political and economic, for the most part are capable only of dealing with relatively short-term objectives, and when put under pressure they barely manage to limp from crisis to crisis.

At the level of international relations the conflicting interests of nation-states, as manifested in their struggle for the economic and political viability of their different institutional structures, overawe common environmental interests and put the latter at the mercy of the short-term considerations of power politics. The domestic politics of the industrialized nations are a miniature of that larger scene. Whether the régime is state-capitalist or state-socialist, the maintenance of economic production constitutes the linchpin of political power, national defence, and international domination. Thus the prevailing wisdom can tolerate a concern for environmental degradation only in so far as it does not affect too adversely the sphere of production.[18] As mentioned earlier, where pollution standards have been instituted, the beginning of economic difficulties brings an immediate cry for the relaxation or lifting of those standards.

Neither in international nor in domestic politics are our existing institutions capable of responding effectively to the dangers of long-range environmental degradation. The manifest inadequacies on the international level are so severe that in all probability nothing short of a crisis situation will force the contending parties to cede jurisdiction and powers of enforcement to an international organization. With reference to the domestic situation in Western Europe and North America, political parties compete principally for the honour of managing troubled national economies, in a setting where few dare yet mention the option of the conserver society. But at least there the existing levels of production could be stabilized and readjusted to different requirements while providing a decent standard of living for the population. In other words, we still have sufficient flexibility in our economic and political institutions to permit us to entertain alternative possibilities, although it would be unwise to postpone the inception of our reconsiderations too long. No particular group of nations, not even the industrialized group, can any longer alter the international situation unilaterally; but the accumulated wealth of the richer ones could cushion the shock of transition to a different perspective on the satisfaction of needs within the domain of the domestic economy.

We cannot now conceive of something like an ethic of biotic diversity being applied on a global scale. However, we can begin to apply the basic principle - a reinterpretation of human needs in relation to the needs of non-human nature - on a smaller scale in the domestic economy. What is urgently required is a beginning, not an illusory leap of faith made in the expectation that an instant and definitive solution is possible. It would be foolish to underestimate the depth of the transformation that is involved in wrenching our perspectives so that we no longer remain indifferent to the needs of non-human nature. In my opinion it is reasonable to believe, however, that the transformation must occur soon and that our immediate task is to search for practical settings in which public awareness of this issue might be aroused.

Where to make a beginning? What is at stake is the recognition that environmental degradation threatens the common interests of many life-forms: it is in this sense that I have sought to relate the needs of non-human nature to the pursuit of satisfaction for human needs. For to have needs is to have interests which the organism or social group must defend in order to survive; and among the range of existing things there exist simultaneously both common interests (or needs) and competing interests. In contemporary society the definition of common interests and the adjudication of competing claims are the province of the legal system. Its basis is the idea of the equality of rights. Until now only human beings and their creations (institutions and artifacts) have been considered capable of exercising rights and therefore of defending their interests. A suggestion has been made recently, however, that we might recognize legal rights for natural entities and by so doing include the interests of the environment and its component parts in the legal processes whereby competing claims are adjudicated.

The author of this suggestion, Christopher Stone, readily admits that the idea will be regarded as unthinkable by most people. But he maintains that the existing legal system (at least in the United States) already has a framework that could be adjusted so as to include this new function: 'It is not inevitable, nor is it wise, that natural objects should have no rights to seek redress in their own behalf. It is no answer to say that streams and forests cannot have standing because streams and forests cannot speak. Corporations cannot speak either; nor can states, estates, infants, incompetents, municipalities or universities. Lawyers speak for them, as they customarily do for the ordinary citizen with legal problems. One ought, I think, to handle the legal problems of natural objects as one does the problems of legal incompetents ... On a parity of reasoning, we should have a system in which, when a friend of a natural object perceives it to be endangered, he can apply to a court for the creation of a guardianship.'[19] Under the law of guardianship a natural entity (a living thing, group of things, or part of terrain) could be represented in a regular legal proceeding, where the present or projected human impact on that entity would have to be explicitly defended and assessed. The right of human agents to appropriate individually or collectively the environment and its resources would then have to be balanced against the right of natural entities to enjoy a certain degree of protection against the pressures of human demands.

Would such a situation mean that humans could not use anything at all in nature? The question itself reveals a misconception about what is at stake in a discussion of legal standing. For the mere fact that an entity (human or non-human) has standing in law does not entail for it an absolute right of existence as against the claims of others; human beings have standing, but they do not possess absolute rights with respect to either their lives or their property, and both may be caused to be forfeited under certain circumstances. What conferral of standing does is to place an entity under the aegis of common procedural safeguards, administered by an authorized tribunal, which specify the means available for the protection of its interests (the right to a hearing in court, the right to sue and collect damages, and so forth). It admits the entity as a full member in that forum where the rights and duties of contending parties are balanced in accordance with common standards of due process. Thus, if this suggestion were adopted, a court could appoint a legal guardian for a polluted lake; the guardian could bring suit against those responsible for the pollution, demanding abatement of it and appropriate penalties; and if damages were assessed the monies would be applied to the task of removing the pollutants or installing protective systems to prevent further degradation.

The hypothesis that everything in nature should be considered as possessing rights and thus standing in law could provide one of the bases for a reference-framework within which human interests could be checked and measured against the interests of other natural entities in the biosphere. In my view it is a particularly appropriate suggestion precisely because our society prides itself on the achievements of its legal code as a fair and just forum for the rational resolution of conflicting interests.

To determine whether the suggestion is at all feasible one must identify and respond to some of the major objections that might be raised against it. First and foremost is the argument that humans alone (and their institutional creations) can have interests and purposes because only humans can articulate them. But is this an adequate response in the light of existing circumstances? Our own scientific understanding has discovered and explicated some of the processes whereby living species and ecosystems such as lakes maintain their integrity as biological entities - what in common language we might call their health. To be sure, this does not imply that there is a pristine, stable condition for any natural entity. Like individual organisms species and ecological subsystems have life-cycles (extending over hundreds of thousands of years) in which they emerge and disappear or undergo radical transformations quite apart from human intervention. We are steadily accumulating more complete knowledge about the phases of these life-cycles, including the stage which their evolutionary course has reached at present. Thus by means of our scientific understanding we could estimate the conditions under which a species or ecosystem would be capable of further development along its evolutionary course, and perhaps use this knowledge as the standard against which to measure the extent of the deviations resulting from human intervention. Obviously our objective would not be to forbid any interventions at all, but rather to establish a scientific conception of the independent integrity of natural things as one side in a contest (the other side being human demands), so that we might have a rational basis for attempting to regulate the claims of competing interests in the biosphere.

This proposal suggests one possible way of creating a new form of awareness about the environmental impact of our needs. It has the following advantages: it relies upon an established forum for the adjudication of conflict; it is consistent with our mode of scientific knowledge and its perspective on nature; it is dynamic rather than static, since it would respond to changes in our scientific understanding. The last point is especially significant, because the open-ended character of our legal system can incorporate changes in knowledge and sensitivity. Our law is based upon the concept of rights; but the kinds of rights that may be exercised at any time, as well as the particular entities (human and non-human) which may exercise them, have varied considerably according to social values and the state of knowledge. The extension of certain previously denied rights to women and blacks is an obvious instance of change with respect to the first variable, and similar changes regarding the second are no less important. For example, present law sets permissible standards for lead levels in parts of the environment, based upon the current scientific understanding of human physiology; these standards may be modified as a result of more advanced knowledge about human health requirements. What we are increasingly inclined to do for the sake of human health – namely, to assess the unintended consequences of production and consumption activities and make the necessary adjustments – we can and must do for the health of the biosphere as well.

Thus the idea of legal rights for natural entities could be grounded in the developing state of scientific knowledge about the functioning of ecosystems. For the knowledge itself is socially impotent; in order for it to play a role in the shaping of behaviour in everyday life, there must be an operational framework whereby this knowledge can be brought to bear forcefully in the institutional processes of society. This present suggestion should not be interpreted as offering an easy solution to the problems inherent in the very logic of a revolutionary social practice (designated in shorthand form as the conquest of nature) that has altered the face of the globe and the ecological situation of the human species over a period of three centuries. Christopher Stone remarks that the idea of legal standing for natural entities is not an end in itself, but rather only one element in an entire series of intellectual, emotional, and institutional changes that would reflect 'a radical new conception of man's relationship to the rest of nature.'[20] The reason for choosing the sphere of law as a reference-framework for these changes lies in the fact that it is an established forum for rational debate and the resolution of conflicting interests; moreover, it has established procedures for translating decisions into effective action. At present the most urgent requirement is a forum for rational debate and action, not simply an abstract statement of new values. We need a practical setting for opening discussion on the initial steps we might take in order to structure differently our world of action.

I have argued that the pursuit of enlightened self-interest should lead us to recognize the needs - and therefore the interests - of non-human nature. In other words, we should attempt to attune our own needs and interests to those of the environmental setting in which we live. One must concede that it is by no means clear at this time that the recognition of legal rights for natural entities is either a feasible or a desirable means to that end. It is possible that such a procedure may be defective in principle or self-defeating in practice, thus creating further confusion with regard to the requirements for a new orientation to our needs.[21] Obviously the existing legal system is much more than a forum for rational debate. It is also an institution like all others whose practices reflect in part the prevailing distribution of social and economic power, which means in effect that its practices often sustain the predominance of particular over common interests. Neither the judiciary nor the legislature is or can be insulated from the distribution of particular interests in the larger social setting; judicial procedures and legislative enactments - particularly the noble pronouncements already enshrined in various environmental protection statutes - can be effectively nullified by the opposition of powerful vested interests. As far as environmental standards for pollutants are concerned, for instance, the promulgation of the standards is one thing, whereas the enforcement of them is quite another. Should this general situation create insuperable difficulties in attempting to implement the idea of legal rights for natural entities, then some other more appropriate procedure must be found for applying the principle of enlightened self-interest.

In some social programs involving the preservation of wilderness and wildlife our present concerns have already carried us beyond the strict confines of short-term, unenlightened self-interest. There is already some public support for overriding the economic calculus in certain circumstances and defending the continuing viability of other life-forms for their own sake. In these instances there is a process at work in which people transcend the market orientation that encourages them to regard the rest of nature as nothing but a means for the satisfaction of human needs. In so doing they begin to define themselves, their needs, and their relationships to each other in a different way. This could be one of the positive alternatives to emerge in the process of our turning away from the high-intensity market setting, where every experience is filtered through the sphere of material objects. Laurence Tribe asks us to imagine a community debate on the merits of building a dam where the participants wish to balance their interests in outdoor recreation against the interests of the wildlife populations in the affected area: 'In choosing whether or what to build - in deciding what technologies to adopt - the community does more than generate a distribution of payoffs and penalties to its members; it also alters irreversibly the experiences available to them (and hence the development of their preferences and consciousness over time) and defines, to a greater or lesser degree, the community's relations with the natural order and its character as a society of persons.'[22] The outcome of the debate in any particular instance would be far less important than the fact that it had been held at all. For in our social activity we *can* choose to nurture modes of experience in our relation to the rest of nature that would begin to carry us beyond the boundaries of enlightened self-interest.

Other satisfactions?

In this essay I have sought to develop a perspective on human needs that is relevant to a specific social situation, namely the high-intensity market setting of contemporary industrialized society in general and in particular of its most advanced representatives. Anthropological, historical, ecological, and other considerations have been marshalled in an effort to uncover the hidden dynamic of that specific situation. This perspective is relevant to the social problems of other contemporary systems only in so far as the commanding position of the industrialized world constrains their own options. For the sake of clarity I have simplified the analysis by ignoring the substantial differences in the levels of development of the market setting that exist across the range of industrial nations and in the variants of state capitalism and state socialism.

Another kind of simplification is involved in the use of aggregate data for my reflections on the environmental impact of the material demands arising in this setting. This essay deliberately ignores the persisting, radical inequalities in the distribution of material benefits and the social injustices associated with them. Those injustices have been amply documented by many other writers. I believe that my approach can be helpful notwithstanding those acknowledged simplifications. The aggregate demands (both current and projected), not the disproportionate shares of the benefits, determine the environmental impact of this system of human needs. A series of potential crises threatens to materialize irrespective of whether or not the extent of the prevailing inequities can be reduced. In this context the most important point is that no effective opposition has been mounted against the ideals (as opposed to the relative injustices) of this setting during the past few decades. Certainly the injustices must be challenged; but no less urgent is the necessity for challenging those ideals which orient all expectations of satisfaction toward the realm of commodities.

Some of the most influential currents of radical thought have borne the implicit message that achieving social justice among the members of human society must precede any attempt to confront the problems inherent in the relation of mankind to the rest of nature. Whether this attitude was right or wrong in principle is now a moot point. Both the inherited antagonisms of social classes and the structural inequities of the high-consumption lifestyle in the industrialized nations are now incorporated in a wider set of contradictions, one that includes as another integral component the interaction of society as a whole with the natural environment. The two dimensions - the antagonisms among forms of human interests and the antagonism between society and environment - are thoroughly interwoven within a contradictory totality. As I have already said, governments, corporations, and some labour unions, for example, have responded to economic difficulties in particular regions or industries with demands for restrictions on environmental pollutants to be relaxed. In such cases the different interests of the three are submerged in the common concern for keeping the economic system on its present course.[23] With few exceptions the most influential social interests support the basic dogma that the sphere of material production must continually expand.

To do nothing more than reject that dogma would be pointless and self-defeating. If the sphere of production simply ceased to grow, and were this to occur in the absence of any restructuring of the modes of satisfaction, the most likely result would be to freeze the present levels of injustice and frustration or even to exacerbate them. It is hard to imagine how a more enlightened social practice could emerge under these conditions. A stabilization or lowering of material production is not an end in itself, nor would it guarantee the coming of a better society. Our essential task is to free the processes for the satisfaction of needs from their tendency to become exclusively oriented toward the blandishments of the marketplace. The first priority of social change in the industrialized nations is to initiate this reinterpretation of needs and satisfaction, regardless of whether the present economy sustains growth or eventually enters a period of prolonged decline.

I have noted at various points that the social practice that tends to orient all needs toward the realm of commodities may emerge in types of industrialized societies which are nominally socialist, as well as in advanced capitalist societies. Similar forces can be detected at work in different forms of the high-intensity market setting when the problem of needs is discussed at the level of abstraction maintained throughout this essay. On the more mundane levels of social policy, however, the similarities are less relevant than the differences, with the exception of those aspects of global environmental degradation (such as the pollution of the atmosphere and the oceans) where the practices of any society, regardless of its economic system or level of industrial development, affect the common interests of many life-forms. There would be little cause for rejoicing in the future realization of the common nineteenth-century utopian vision - a peaceful association of co-operative communities around the globe - if these communities were to inherit the wasted and despoiled remains of a once-bounteous planet.

On the level of social policy one must take full account of the different sets of options for change and institutional constraints in the nations of Western Europe, Japan, and North America, for example, where formal political democracy coexists with the economic régime of multinational corporations, as opposed to Eastern Europe and the Soviet Union. The possibilities for social change which were included under the rubric of the conserver society refer to the situation of the wealthier present-day capitalist societies. I do not know whether the thread of argument in this essay is at all relevant at the level of practical social change to the state of affairs in the industrialized socialist nations, or in China, or elsewhere. It *is* relevant to the situation of those (especially the wealthier representatives of the capitalist group) which already possess a level of material production sufficient to enable them to begin diverting a considerable portion of those productive resources to other ends - namely, the encouragement of alternative routes to satisfaction. The existing capacities for material production in these countries supply a far wider range of options than is available to others. And it is with reference to the range of possible options in the wealthier industrialized nations that, I suggest, we can begin to adopt a new orientation to both the social and the environmental dimensions of our needs.

I wish to emphasize the point that these are only possibilities. Whether the institutional arrangements of our society can adjust to the requirements of this program through relatively peaceful transitional stages is an open question. Nor is there any assurance that the majority of the population would welcome an opportunity to restructure the modes of satisfaction accordingly, since this would be indeed a radical departure from the prevailing way. What is in question, then, is both the willingness and the ability to reorient our behaviour, our institutions, and our technology toward a social practice that would find sources of satisfaction in curtailing the upward spiral of industrial production, in reducing the senseless squander of resources and energy that fuels the ambiguous dialectic of needs and commodities.

An assertion that this adjustment is beyond our capabilities would be one of the most devastating judgments on the prevailing system that could be made. It would contradict some of our most cherished beliefs about the superiority over earlier societies of a society based on sustained technological progress. First in rank of the claims made on its behalf is that the technologically based society opens a range of possibilities far broader than anything hitherto imaginable.[24] But does the spectrum include the possibility of freeing ourselves from dependence on unlimited economic growth? Since the question has not yet been posed in the practical realms of economics and politics, no one is now in a position to answer it. The necessary preliminary exercise of clarifying the precise nature of our dependence on endlessly expanding production has not yet been completed. In this essay I have attempted to understand this dependence as a function of a social system which tends to narrow the range of possibilities radically, to inhibit all modes of satisfaction except that of consumption. Just because the possible modes of satisfaction have been narrowed so drastically, the dimensions of the one favoured activity grow uncontrollably like a cancerous cell. The decisive future test for this society's institutions is whether or not they can muster public support to contain the cancer and open alternative possibilities.

The critique of the expanded market setting in terms of its deleterious consequences for society and individuals has its roots in the classic contributions of social thinkers such as Marx, Veblen, and Durkheim. Its subsequent transformation since their time into what I have called a high-intensity market setting has revealed its implications for the relation between man and nature - a dimension that has been present from the beginning, but has remained largely unrecognized until now. From our present vantage-point, however, we can see that the social and environmental dimensions of our situation are so fully interwoven that it would be fruitless and inappropriate to isolate one from the other. We can and should confront the two aspects simultaneously rather than serially. Attempting to transcend our present orientation means to encourage changes in our sense of satisfaction and well-being and also to encourage a respect for the 'independence' of non-human nature.[25]

Respecting the independence of non-human nature - regarding it not only as a source of human satisfaction but also as an end in itself - involves nothing particularly mysterious. In beginning to reorient our perspectives in this way we need only make a consistent effort to consider far more carefully the character of our objectives in transforming and appropriating nature for our use. We should think more carefully about the quality of the experiences that we are denying to ourselves, as well as those that we wish to make available, when we organize our relationship to the natural environment so exclusively around the hunt for resources.[26] In accordance with the 'negative theory of needs' mentioned earlier, this is a 'negative' conception. Respect for the independence of non-human nature means that we would not force the entire spectrum of possible experience arising out of our interaction with nature into a single mould.

'Respect for nature' is a metaphorical expression, just as is the more familiar phrase 'domination over nature'. It should not be forced to carry too precise a definition; interpreted in a metaphorical way, however, it may usefully serve as an image to guide the reorientation of our behaviour and attitudes. I should like to suggest first how *not* to interpret it, and then outline what I think is an acceptable version of it.

'Nature knows best' : this timeworn phrase is about as good a summary of conventional wisdom as any. In this form respect for nature means either not interfering at all with so-called 'natural' processes such as diseases, or only using 'natural' substances as opposed to industrially-created chemicals. There are two basic errors in this attitude. In the first place, the very existence of human social organization represents a fundamental 'interference' with the natural processes that make up the life-cycles of all other creatures. Human cultures transmit a far more elaborate collection of learned behaviour patterns through successive generations than do other social animals, and these patterns condition their responses to fluctuations in environmental conditions. Thus even in primitive societies human population and health fluctuations are in no way comparable to predator-prey cycles, for instance, among other animals. To put the point crudely : we never obey nature, but rather our conception of nature.

Secondly, the naïve preference for natural instead of industrially-produced substances – common among some of the 'health-food' devotees – is based on an inadequate knowledge of chemistry. Traditional pre-industrial techniques, including the uses of plant materials, are based upon chemical and biological reactions learned through trial-and-error experiments rather than laboratory analyses; some are indeed both appropriate and beneficial, some have neutral consequences, and some we know to be downright harmful. Our industrial techniques likewise produce results ranging from the beneficial to the harmful. The point is that one cannot distinguish the two different sets of impacts using the categories of natural vs. unnatural or artificial. Finer discriminations are necessary. One can reasonably prefer vegetable over aniline dyes, for example, for at least two different reasons : they yield more pleasing colours (according to some tastes) and they are far more benign, in terms of their environmental side effects, as we produce and consume them. One can also reasonably prefer the taste of meat from animals who have been raised in traditional ways, as opposed to those who have been dosed from birth with hormones and tranquillizers in overcrowded pens or feed-lots; or the humble backyard tomato's taste over its recent rival's, the fruit of genetic engineering designed to produce a shape and skin thickness suitable for mechanical pickers. But one can also appreciate the quite sensible improvements in sanitation, comfort, and convenience wrought by industrial technology that are genuine contributions to our well-being.

In a very real sense our industrial technology's chemical wizardry is out of control. It tempts us with innumerable trivial novelties; it degrades the quality of our preferences, for example by presenting bizarre laboratory concoctions to us and calling them foods; it is shockingly careless with toxic substances like the dioxin which showered the Italian town of Seveso; and it has not the slightest idea what the long-range synergistic effects of all these apparently clever techniques may be. As willing consumers of so much junk we co-operate with this irresponsible profit-seeking – although it must be said that most people are utterly unaware of the risks we run. So long as we prefer to accept without proper scrutiny whatever appears on the store shelves, the structure of oligopolistic market competition will require all producers to take part in provisioning the chemical feast for us.

The dimension of possible risks is unknown. I believe it would be prudent to shrink it by bringing our industrial technology under control, by gradually forbidding all inessential chemical applications and by requiring far more stringent laboratory tests of long-range environmental impacts before any chemical substance is approved for general use. Obviously what is deemed essential or inessential in consumer goods is a matter of judgment, and a more prudent course of action will depend on building citizens' support for more enlightened public policy. A proposal to better control our technology is, alas, not as dramatic as a call to reject it altogether, and at the present time it is not likely to fare well in the competition for public attention. It is disheartening to surmise that nothing short of successive environmental crises on the model of the Seveso incident would be sufficient to drive home the point.

Respect for nature should not be based on a blanket rejection of science-based technology and a fallacious belief that following the natural course of events is the best guideline for our behaviour. Equally as fallacious, I have suggested, is the view that we can expect our science and technology eventually to achieve such mastery over natural processes that we shall be able to pursue whatever whim happens to overtake us. But is there a middle ground? I believe so. Our scientific culture can entertain a rational conception of respect for nature, and this in turn can assist the larger culture based on our industrial technology to modify its course in practical affairs.

Modern science allows us a better understanding of *how* natural processes occur than other natural philosophies do. While we should not regard this as the only valid interpretation of what nature signifies to us, it will be and should be one significant mode of our relationship with non-human nature from now on. We understand much better than in the past the immense biological and chemical complexity of our physical environment and the species that inhabit it. We can make good use of this knowledge; but we need not assume that we can or should 'control' those complex interactions fully. We *could* conclude, on the basis of our scientific knowledge, that there are far too many risks involved in attempting to manipulate larger and larger sets of ecological interactions for the sake of short-run benefits. Our social institutions have real limitations, owing to their susceptibility to conflicting interests and outbreaks of violent conflict, and to the difficulties of assimilating complex information sources in public decision-making processes. These limitations govern the practical impact of scientific knowledge on social, economic, and political choices. As we raise the levels at which we attempt to manipulate ecological interactions, we increase enormously the number of variables that must be controlled, as well as the magnitude of the possible negative externalities that may appear. We may or may not have adequate theoretical models for controlling the variables at any time; but even if we do, there is always the danger of 'overloading' the practical capacity of our social institutions to carry out the necessary measures.

Scientific knowledge, as it achieves a more comprehensive grasp of nature's complexity, may help us to realize the *practical* impossibility of manipulating large-scale ecological interactions. If it does, and if we accept the restraint, this would be an example of what I mean by a rational conception of respect for nature. It does not deny that we possess the 'power' – the scientific and technological capacity – to undertake those operations; it means that, with full knowledge that we do have the necessary instruments, we recognized the limitations of *our* nature and our social institutions, and chose not to do so.

Our cherished position as the earth's dominant species would hardly suffer as a result. We would still cover the planet with our settlements and continue to modify or destroy the habitats of other species, and some of the species themselves as well. This kind of respect for nature, the renunciation of large-scale environmental manipulations, would be undertaken in the first instance as the most prudent course for the long-term preservation of our own well-being. It would also have appreciable 'spinoff' benefits for other living entities, by reducing both the direct and indirect pressures we exert on them. This would be at first only an unintended consequence of a more enlightened pursuit of self-interest for ourselves; but in time we might find that seeing ourselves as members of a community of living entities, rather than as unchallengeable masters of a conquered domain, yields a finer sense of contentment and well-being than we have been able to find on our present heading.

There is an old idea of man as the 'steward' of nature in the Christian tradition. A secular version of this idea may be an appropriate adjunct to the rational conception of respect for nature. We have already modified most of the habitats on this planet, and we are well on our way toward completing the process; we are in a position to control or eliminate many (or perhaps any) other species through predation, modification of habitats, or genetic manipulation. In a very real sense, therefore, we are 'responsible' for the fate of all living things. For some persons it would be sufficient to reserve a few specimens of those that appeal to us in zoos and to let the rest take their chances. This is a rather limited notion of stewardship. A more generous interpretation would be to preserve as much as possible of the remaining habitats, with suitable compensation to those nations and groups whose economic interests are injured as a consequence.

Respect for nature as the exercise of voluntary restraint on our operational capacity to manipulate the environment has another aspect: respect for the full range of possible human experience in our relation to nature. The ever greater predominance of seeking operational control over nature in our culture restricts our experience to the dimension of matter-energy transformations that are expressed as 'laws' of governing natural processes. But the numbers and material demands of the species grows at an equal or faster rate, always renewing the violent struggles over the spoils of dominion. No amount of operational control can resolve this dilemma; any hope of doing so resides only in a reordering of social relations. The industrialized nations should have accumulated enough evidence by now to realize that the existing political economy generates fresh material demands with each new level of production; the escalating demands intensify the search for further advances in operational techniques. Other dimensions of possible experience are repressed under the weight of this self-defeating enterprise. To understand the necessity for restraining our operational capacities is to open the way toward a fuller and more balanced experience of nature, one that does justice to its sensuous or aesthetic qualities.

Respect for nature really means respect for our own nature, for the full range of experience of which it is capable – and for its limitations as well. It is simply not possible (even if it were desirable) for our social institutions to manage with sufficient care the full effects of large-scale ecological manipulations. We must recognize those limitations and try to live with them.

Admittedly this is a vague formulation. But there is no positive program that can be outlined for this purpose, just as there cannot be a detailed program for alternative modes of satisfaction. My primary objective in this essay has been to mark more clearly the dangers in the terrain which we have been traversing with such confidence. I have contended that the prevailing understanding of our needs and their relation to a specific form of satisfaction prevents us from recognizing those dangers. I have not presumed to offer solutions, but only a prognosis based upon the examination.

It would be fatuous for an author to suggest that he had written the prescription when the accuracy of the diagnosis had not yet been agreed upon by those who are supposed to benefit from the effort. Agnes Heller remarks: 'In the free market system, the theory comes to the free market, too: People take it or leave it, make use of it or neglect it according to their own needs.'[27] In the marketplace of ideas, as in current medical practice, prescriptions are not hard to obtain. The reader anticipates that the diagnosis will be followed by the cure, and many would rather risk being given a placebo than go away empty-handed. Most current commentators on the ills of modern society obligingly furnish a remedy of some sort.[28] Imaginative authors sometimes compound a cure out of exotic materials drawn from distant cultures, such as ancient Buddhism or the ways of the native American tribes, and in entertaining these the reader need not perceive any immediate threat to his established ways.

The redirection of social change away from the high-intensity market setting would signify a collective willingness to confront the problem of human satisfaction in ways not now considered feasible or desirable. It would initially signify only a willingness to recognize *how* and *why* we are increasingly confused about our needs and to understand that in this setting we tend to regard non-human nature only as a means for our ends. Any prescription is illusory if it suggests that it can guide us to a point beyond enlightened self-interest where we would realize, suddenly and completely as in a vision, what is involved in regarding the rest of nature as not only a means but also an end in itself. A beginning in this regard may be no more dramatic than a serious public commitment to the preservation of endangered species, the designation of wildlife areas that have strict constraints on human uses, and a drastic reduction in the types and amounts of toxic wastes discharged into the environment.

We too can come to understand, as John Stuart Mill did long ago, that accepting the limits to growth can be the catalyst for new forms of creativity and fulfilment in our culture. But we can do so only by recognizing the limits to satisfaction that we have imposed on ourselves by restricting our search for well-being so severely. To bind all our hopes to the hunt for resources with which to continue feeding our industrial technology is pointless and self-defeating. Indeed, it has all the markings of a massive social obsession whose destructive aspects can only worsen with every further achievement in extending the interplay of needs and commodities.

The abundance we have created is deceptive, in that we cannot perceive the depth of our dissatisfaction with the social forms through which we produce and consume it. We need only to begin changing those forms in order to release the immense potentialities for individual self-fulfilment that already exist in the world around us. The possibilities for satisfaction that might be drawn from different forms of productive activity and of our relationship to non-human nature, and that are now so deeply suppressed, can minister to our needs far more effectively than can any new assortment of goods. In orienting ourselves toward these suppressed possibilities we can discover some of the abundant sources of satisfaction that have lain untapped for so long.

Notes

Full bibliographical information for publications cited here will be found in the list of works cited (pp. 143-51).

INTRODUCTION

1 Sources for the statistics given in the preface are as follows: per capita U.S. consumption of materials and projections for the year 2000, *Material Needs and the Environment* figures 2.3, 2.6, 2.13, 2.22; share of world consumption of resources by the developed nations, ibid p. 9-5; U.S. Bureau of Labor Statistics report cited in Parker *The Myth of the Middle Class* 15; report by the International Economic Policy Association, a research group sponsored by U.S. business firms, cited in the *Globe and Mail* (Toronto), 22 July 1974.

2 Manning Nash 'The Organization of Economic Life' in Dalton, ed *Tribal and Peasant Economics* 9; Herskovits *Economic Anthropology* 58

3 For a recent report on the index, which was developed by the University of Michigan's Survey Research Center, see Mandell et al *Surveys of Consumers, 1971-72.*

4 An exponential use index for nineteen non-renewable resources, calculating potentially available resources at five times the currently known reserves, ranges between 29 years' supply of gold and 173 years' supply of iron, with the median around 60: Meadows et al *The Limits to Growth* table 4. (For a criticism see R.W. Page's article in H.S.D. Cole et al *Thinking About the Future*.) A recent study sponsored by the U.S. National Academy of Sciences, *Mineral Resources and the Environment*, predicts the coming of serious shortages within a few years if conservation policies are not adopted (*New York Times* 12 February 1975).

5 Landsberg et al *Resources in America's Future* 427 ff

6 *Material Needs and the Environment* table 9.1 and p. 9-5; Parker *The Myth of the Middle Class* 212
7 André Gorz *Socialism and Revolution* 198; italics in original
8 Mill *Principles of Political Economy* book 4: chapter 6, Marx *Capital* 1: chapter 1, Veblen *The Theory of the Leisure Class*, Durkheim *Suicide* 246-58, Lukács *History and Class Consciousness* (essay on reification), Tawney *The Acquisitive Society*; among more recent writers: Polanyi *The Great Transformation*, Fromm *The Sane Society*, Galbraith *The Affluent Society*, Macpherson *The Political Theory of Possessive Individualism*, Horkheimer and Adorno *Dialectic of Enlightenment* ('The Culture Industry'), Marcuse *An Essay on Liberation*, Baran and Sweezy *Monopoly Capital*, Linder *The Harried Leisure Class*, Illich *Tools for Conviviality*, Henry *Culture against Man*, Slater *The Pursuit of Loneliness*, Weisskopf *Alienation and Economics*.

In recent European social thought some important contributions are: Debord *Society of the Spectacle*, Baudrillard *La Société de consommation* and other books. Giradin 'Toward a Politics of Signs: Reading Baudrillard' is a brief introduction to the latter. Also: Haug *Kritik der Warenästhetik*, Menge *Der Verkaufte Käufer: Die Manipulation der Konsumgesellschaft*, Schneider 'Neurosis and Class Struggle: Toward a Pathology of Capitalist Commodity Society.'

9 For example, Mishan *The Costs of Economic Growth*, Schumacher *Small is Beautiful*, Goulet *The Cruel Choice*, Heilbroner *An Inquiry into the Human Prospect*, Meadows et al *The Limits to Growth*, Dator 'Neither There nor Then'; for background reading see two collections: Thomas, ed *Man's Role in Changing the Face of the Earth* and Farvar and Milton, eds *The Careless Technology*.
10 For example, Commoner *The Closing Circle*, Ehrlich and Ehrlich *Population, Resources, Environment*, Falk *This Endangered Planet*, Passmore *Man's Responsibility for Nature*, essays by Kenneth Boulding, Rodman 'The Ecology of Openness,' Dubos *So Human an Animal*, Eiseley *The Invisible Pyramid*, Livingston *One Cosmic Instant*, Shepard *The Tender Carnivore*

PART ONE

1 Twedt 'Consumer Psychology' 280. For a description of other techniques for measuring responses to product messages (such as pupil dilation) see Perloff 'Consumer Analysis' 445-6, 455.
2 Twedt 265. A useful collection of articles is Gaedecke and Etcheson *Consumerism*. A recent radical perspective will be found in Fritsch *The Contrasumers*.
3 Katona 'What is Consumer Psychology?' 226

4 See Scherf 'Consumer Education as a Means of Alleviating Dissatisfaction' for a discussion of the problem of satisfaction in consumer behaviour.
5 *Montreal Gazette* 31 December 1973
6 Recently the U.S. Office of Education completed a four-year study of literacy in the adult population of the United States, including a special section on consumer economics. The study found that 30% were functionally illiterate in tests involving the understanding of newspaper advertisements and an additional 33% were functional but not proficient in such tests (reported in the *Globe and Mail*, 30 October 1975).
7 Hettich 'Consumption Benefits from Education' 197. For an excellent discussion of the problem of knowledge about commodities see Scitovsky *Papers on Welfare and Growth* 193ff.
8 A biochemist estimates that Canadians ingest about 2,500 commercially produced chemicals each day and remarks that no research has been done on the combined effects of them. There are an estimated 500,000 chemical formulations available in over-the-counter drugs (*Globe and Mail* 16 October 1975).
9 Zwerdling 'Death for Dinner' (hyperactivity, nutrition), *New York Times* 16 July 1974 (carcinogens), *Globe and Mail* 11 July 1974 (cosmetics). Another dimension of this problem is exposure to chemicals in the workplace: Stellman and Daum *Work is Dangerous to your Health* 155. There are 600,000 chemicals used in industry, and several thousand new ones are introduced every year (*Globe and Mail* 13 February 1975).
10 Silverman and Lee *Pills, Profits and Politics*
11 Twedt 'Consumer Psychology' 272, 278-9, 281
12 Linder *The Harried Leisure Class* chapter 2, de Grazia *Of Time, Work, and Leisure*
13 Linder 96; the preceding quotation is from pp. 91-2. A recent critical note by Albert Hirschman, 'An Alternative Explanation of Contemporary Harriedness,' argues that the high-intensity consumption society is a 'passing phase.'
14 The study (by Professor Jaro Mayda) is cited in Falk *This Endangered Planet* 200.
15 Some authors have argued that there is a structural tendency toward a deterioration in the quality of goods in a market economy. For two different versions of this argument see Hollis and Nell *Rational Economic Man* 218 and Hirschman *Exit, Voice, and Loyalty* 47-54.
16 Krieger 'What's Wrong With Plastic Trees?' 451, 453
17 Lipsey et al *Economics* 5. An excellent critical article on this subject is Saul Engelbourg's 'Insatiability, a Problem for Utopia?' Economists distinguish between 'wants in general' and particular wants and assume that only the

former are insatiable. Also, within a specific period of time a want for anything is regarded as being satiable. For recent work utilizing a general satiability assumption see Barnett 'The Effects of Consumer Bliss on Welfare Economics.'

18 Katona et al *Aspirations and Affluence* 171. The research reported in this volume shows some significant cross-cultural differences, regarding the consumer's attitude toward what may be expected from the marketplace, in the United States, Great Britain, France, Holland, and the Federal Republic of Germany.

19 In this discussion of obesity I am relying on an unpublished paper by Lorne Elkin, 'Commodity Consumption and Human Need' (1970). Of course norms for obesity vary widely among different cultures.

20 Katona *Aspirations and Affluence* 13

21 Miller *Economics Today* 38. In an age of lengthy and enforced universal schooling, textbooks provide an indispensable key for understanding the prevailing social consciousness. A more adequate statement of the problem of scarcity is given by Heilbroner *The Making of Economic Society* 17. Since I am dealing only with the 'common-sense' understanding of scarcity, I am ignoring the more technical aspects of this subject, such as Menger's distinction between scarcity and insufficiency. Since most resources have alternative uses (i.e., they can be used to accomplish different goals), one always needs criteria for distributing them sensibly for the attainment of an entire set of goals.

22 Sahlins *Stone Age Economics* 4. Cf Stanley 'Nature, Culture, and Scarcity.' Sahlins' essay is in part a polemical contribution to the ongoing debate concerning formalism vs. substantivism in economic anthropology. The concepts of insatiability, scarcity, and economic surplus have been drawn into this debate. See Dalton *Economic Anthropology and Development* 47-8, 202-16, and the references cited there.

23 The phrase *psychology of scarcity* is used by Etzioni (*The Active Society* 618).

24 In New York City in 1906 'horsedrawn traffic averaged 11.5 mph in the city center; by 1966 it was crawling at 8.5 mph, drawn by the world's most powerful mass-produced engines' (Rivers *The Restless Generation* 73).

25 Data on estimated 1970 and projected 2000 amounts will be found in Brubaker *To Live on Earth* table 3.

26 Kneese et al *Economics and the Environment* 7. Cf. Victor *Pollution: Economy and Environment* 18-19.

27 Mishan *Making the World Safe* 17

28 *Globe and Mail* 10 January 1975

29 See, for example, Harris *Culture, Man, and Nature passim* and Vayda, ed *Environment and Cultural Behavior.*
30 I have undertaken a critical study of this notion in a previous book, *The Domination of Nature.*
31 See Passmore *Man's Responsibility for Nature* chapters 1 and 2.
32 *Works* 4:115. See further Leiss *The Domination of Nature* 48-57, 190-8.
33 Bacon's phrase is in *Works* 3: 302.
34 See *The Philosophy of Francis Bacon* 62.
35 Macpherson *Democratic Theory* 17-23, 29-36, and *passim*
36 *Grundrisse* 410
37 A discussion of this point will be found in Leiss *The Domination of Nature* 178-90.
38 *The Philosophy of Francis Bacon* 92-3
39 *Memories and Studies* 291. John Rodman cites Descartes' remark that the concept of animals as automata enabled man to kill animals without a sense of guilt ('The Dolphin Papers' 24).
40 'What's Wrong with Plastic Trees?' 451, 453
41 'Ways Not to Think about Plastic Trees' 1347

PART TWO

1 *Principles of Economics* 1: 92, footnote. For an account of subsequent discussions on the measurement of 'utility' see Kauder *A History of Marginal Utility Theory* 191 ff.
2 Bradburn and Caplovitz *Reports on Happiness* 56-7
3 Rescher *Welfare* 45
4 *Freedom and Culture* 72
5 'Culture as a Determinant of Behavior' 160
6 Etzioni recently revived the argument of basic human needs (*The Active Society* 622-6), but he did so for a specific purpose, namely 'to test empirically the key proposition that the flexibility of basic human needs is limited in that they can be more readily and fully satisfied in some societal structures than in others' (624). This is an important proposition, and his analysis of it is to some extent independent of his claim that a set of universal basic needs can be specified.
7 First enunciated in 'A Theory of Human Motivation' and elaborated in *Toward a Psychology of Being.* Much of the empirical research in recent social psychology has employed Maslow's influential scheme.
8 *The Sane Society* 20 (italics in original)
9 See Katona *The Mass Consumption Society* chapter 7 for a forceful presenta-

tion of the behaviourist viewpoint. In *The Affluent Society* and other books Galbraith has developed the notion of artificially created wants.

10 *An Essay on Liberation* chapter 1

11 In addition to the literature already cited, see especially Bay 'Needs, Wants, and Political Legitimacy' and Braybrooke 'Let Needs Diminish that Preferences May Prosper.' In this context it is interesting to compare the older theories of the hierarchy of wants in economic thought with the more recent psychological literature on the hierarchy of needs. For the former see Bowley *Studies in the History of Economic Theory before 1870* 73-80 and Georgescu-Roegen *Analytical Economics* 190-8; and Parsons *The Structure of Social Action* chapter 4.

There is considerable discussion of the nature of wants in recent philosophical literature. For example, W.D. Hudson (*Modern Moral Philosophy* 317) argues the wants-needs distinction by saying that 'anything can (logically) be wanted for its own sake; but nothing can (logically) be needed for its own sake.' However, I am not persuaded that this kind of formal distinction can help to resolve the practical social problems that have been outlined in this essay.

12 Cf. Dumont *La dialectique de l'objet économique* 283: needs are 'the primary intentions around which symbolic structures are developed.'

13 See Sahlins *Stone Age Economics* chapters 2 and 3 for an extensive analysis of the domestic mode of production.

14 *Capital* 1: 72

15 The concept of homeostasis is used in biology with reference to both the activity of individual organisms and the relation of a species to the environment. I am extending the concept by analogy to the planetary biosphere in its relation to the larger 'environment' of outer space. It should be noted that this is a biological or ecological concept and that it cannot be applied to sociocultural formations.

16 Smith *An Inquiry into ... the Wealth of Nations* 13. This anthropocentric conceit has been overcome only recently.

17 Sahlins *Stone Age Economics* 302

18 Godelier *Rationality and Irrationality in Economics* 297 (citing Bohannan's research)

19 *The Psychology of Economics* 20

20 See especially *Contribution to the Critique of Political Economy* chapter 1 and *Grundrisse* 88-100.

21 *Grundrisse* 156-8; italics in original. See generally Polanyi *The Great Transformation.*

22 *Ibid* 496

23 *Ibid* 514-15. See further my article 'Critical Theory and its Future' 343-6.

24 *Grundrisse* 157: In production for exchange 'all individuality and peculiarity are negated and extinguished. This indeed is a condition very different from that in which the individual or the individual member of a family or clan (later, community) directly and naturally reproduces himself ... ' Cf. Moore 'The Metaphysical Argument in Marx's Labour Theory of Value' and 'Marx and the State of Nature.'

25 *Capital* 1: 43, 36, 35 respectively. Of course Marx was not unaware of the dimension of symbolic mediations: 'Firstly, the object is not an object in general, but a specific object which must be consumed in a specific manner, to be mediated in turn by production itself. Hunger is hunger, but the hunger gratified by cooked meat eaten with a knife and fork is a different hunger from that which bolts down raw meat with the aid of hand, nail, and tooth' (*Grundrisse* 92). In this section Marx's main concern was to show the predominance of the form of production over that of consumption, however, and he did not analyze further the symbolic correlates of needs.

26 *Grundrisse* 694

27 *The Theory of Political Economy* 105; italics in original. See Kauder *A History of Marginal Utility Theory*, especially chapter 11. Most recent economics textbooks do not bother to define utility at all, restricting their discussion to an explanation of the principle of marginal utility. Samuelson identifies (in passing) utility and 'psychological' utility. Miller (*Economics Today* 407) says: 'Utility: a general term applied to the satisfaction people subjectively obtain from consuming a product or service.' In the remarks in the text I have ignored the shift from cardinal to ordinal utility in economic theory. See generally Schumpeter *History of Economic Analysis* 1053-66.

28 This non-normative or behaviourist orientation results in a conception of 'neutral' or 'formal' utility, which apparently made the marginal utility theorists uneasy even as they adopted it: ' ... neutral utility has not solved but rather only shelved a question whose answer is important for limiting the field in which marginal calculations are valid. It seems unlikely that the situation can be clarified without studying the relevant parts of psychology. The role of psychology for our whole field can be easily underrated, and the discussion of the law of diminishing utility would have been barren without the advice of the psychologists' (Kauder 134). One further instance: 'The elements of scarcity and choice, which are the outstanding factors in human experience that give economic science its reason for being, rest psychologically on firm ground' (Herskovits *Economic Anthropology* 3). Some economists apparently cling to the pious hope that psychologists will bail them out of

their conceptual difficulties, thus remaining blissfully unaware that the latter in turn look to either sociology or physiology and chemistry for the same kind of succour - a notorious vicious circle. Cf Hollis and Nell *Rational Economic Man* 138-9.

29 Lancaster *Consumer Demand: A New Approach* 116

30 Ibid 114. In an earlier formulation of his theory Lancaster wrote: 'The term "characteristics" was chosen for its normative neutrality; in my earliest draft of this idea I called them "satisfactions," but that has too many connotations' ('Change and Innovation in the Technology of Consumption' 14). Too many connotations indeed!

31 Lancaster 'Change and Innovation in the Technology of Consumption' 21

32 *Consumer Demand* 7

33 Wolin 'Looking for "Reality" ' 17

34 Becker 'A Theory of Marriage'

35 If one took at face value the account by Carlos Castenada of his apprenticeship with the Mexican 'sorcerer' Don Juan, one would conclude that the first (imaginary) situation involves much greater uncertainties than the second. Castenada experienced far more difficulties in entering Don Juan's 'world of objects' - that is, the sorcerer's interpretations of the significance of the local flora, fauna, and topography - than Don Juan did in making use of Castenada's contributions to their association (a sporadic supply of groceries and automobile journeys).

36 Ewen 'Advertising as Social Production: Selling the System.' Cf. Baudrillard *La Société de consommation* 127.

37 Marx has an interesting perspective on this matter. He maintained that one of the progressive traits of capitalism was to undermine 'all traditional, confined, complacent, encrusted satisfactions of present needs, and reproductions of old ways of life,' and he believed that the 'expansion of needs' is a positive aspect of human development (*Grundrisse* 410). The capitalist 'searches for means to spur them [the workers] on to consumption, to give his wares new charms, to inspire them with new needs by constant chatter etc. It is precisely this side of the relation of capital and labor which is an essentially civilizing moment, and on which the historic justification, but also the contemporary power of capital rests' (ibid 287). (Note especially the last phrase in this passage.) These comments are less well known than is his argument that consumption is a subordinate element in the overall social process and that production is the dominant element (ibid 90-4).

Even if one agrees with this last-mentioned argument, however, one must go beyond Marx's formulations. The growing separation of the two spheres in the everyday life-experiences of individuals, brought about by the enormous expansion of the market economy since his time, creates a problematic situation in the sphere of consumption that is distinct from that in the sphere of production. For a different perspective on this point see the two excellent articles by Herbert Gintis, 'Consumer Behavior and the Concept of Sovereignty' and 'A Radical Analysis of Welfare Economics.'

38 According to the prevailing wisdom, it is rational for individuals to choose a small but certain gain at the risk of incurring a large, but highly uncertain, loss. We may have to revise our conception of rational behaviour in this regard with respect to the possibilities of serious environmental degradation.

PART THREE

1 Ehrlich et al *Human Ecology* 206

2 'The most serious population growth occurs among the affluent whites of the USA, and their analogues in Western Europe, the Soviet Union, and Japan. These people are the prime looters and polluters of our planet – the ones who are destroying our life-support systems' (Paul Ehrlich 'Letter to Walter Thompson' in the teachers' manual for *Population, Resources, Environment* 20).

3 Etzioni *The Active Society* 626; Marcuse *One-Dimensional Man* 6

4 *Principles of Political Economy* 2: 756

5 See especially Daly, ed *Toward a Steady-State Economy* and Heilbroner *An Inquiry into the Human Prospect* 82-95.

6 Mill's phrase *the stationary state* unfortunately has certain negative overtones – such as the image of frozen social patterns – which of course Mill himself did not wish to impart. More recent theoretical and practical explorations of alternative patterns have revived the image of a dynamic society motivated by the objectives of resource conservation and greater personal satisfaction in producing activity through institutional changes. These are in my view consistent with what Mill had in mind. In the rest of this section I will use *the conserver society*, which has more positive overtones, as a general designation for the objectives incorporated both in Mill's own notion and in the recent initiatives.

7 Caplovitz *The Poor Pay More* 12-13. See also Moscovici and Columelli 'Contributions à l'étude des aspirations économiques des travailleurs.'

8 Obviously a critical study would indicate the considerable differences in the perspectives of the thinkers mentioned above; here only some of the common features will be mentioned. Such a study would also have to detail the inadequacies in these perspectives and seek to reinterpret their ideals more concretely in the context of our present situation.

9 *Tools for Conviviality* 11, 58

10 Authors such as Mishan have stressed the hidden costs of the prevailing pattern of production. In part the existing pattern became viable only due to cheap energy (a fast-vanishing feature of our lives), for example in mechanized agriculture: see Pimentel et al 'Food Production and the Energy Crisis.'

11 Practical research on these alternatives is now being undertaken by individuals and small groups such as the 'New Alchemists' in Massachusetts and PEI. Some attention and support for these activities recently has been forthcoming from government and industry sources, but of course there remains the vast disproportion between social investment in these alternatives as compared to the massive technologies of the dominant ideal. See, for example, Dickson *Alternative Technology and the Politics of Technical Change.*

12 In this context energy demand refers to our reliance on *all* external sources of energy, including renewable sources such as solar and wind energy. The reasons for minimizing energy use involve both thermodynamic (environmental) and socio-political considerations. For a provocative discussion on this point, which proposes a research program on 'socially optimal energy quanta,' see Illich *Energy and Equity.*

13 See Carpentier 'Organisational Techniques and the Humanisation of Work' for one general account of this matter.

14 Mill *Principles of Political Economy* 2: 756

15 Another statement of this problem from a related perspective may be found in Moscovici *La Société contre nature* chapter 12.

16 *Science* 184 (10 May 1974): 646

17 In the development of the controversy over aerosol propellants and their impact on the atmospheric ozone, some scientists called for an immediate ban on the use of aerosol sprays. See the *New York Times* 26 September 1974, 1; 27 September, 82; 31 October, 29; and 21 November, 29, 46.

18 For a thorough discussion of the international problem see Falk *This Endangered Planet.* As far as the U.S. economy is concerned, a recent study for the Environmental Protection Agency concluded that the pollution controls presently written into U.S. law will have very little long-range impact on economic growth (*Globe and Mail* 11 February 1975).

19 *Should Trees have Standing? Toward Legal Rights for Natural Objects* 17. Only a few aspects of Stone's intriguing argument are presented in the text; the reader is urged to read it in its entirety before attempting to judge the merits of his case.

20 Ibid 48

21 Stone's proposal applies specifically to the U.S. political and legal system. In others (particularly in parliamentary systems) any implementation of the general tenor of his proposal would entail a somewhat different approach.

22 Tribe 'Technology Assessment and the Fourth Discontinuity' 656

23 In order to avoid misunderstanding, I must repeat that I am aware of the social injustices inherent in the present arrangements. I am assuming that these are well known; I also assume there would be general agreement on the point that the injustices will not be made to disappear merely by expanding the sphere of production. It is sometimes said that a concern for environmental degradation is an issue that reflects the interests of the affluent middle class. I regard this not only as a crude and short-sighted judgment, but also as an indication of simple ignorance, for so far those who always have been most directly and seriously affected by environmental degradation are the industrial production workers.

24 For a discussion of this point see my essay 'The False Imperatives of Technology.'

25 The most careful discussion of what is meant by an attempt to see 'beyond enlightened self-interest' in our relation to nature will be found in Marcuse's *Counterrevolution and Revolt* chapter 2.

26 A fine expression of these concerns is Loren Eiseley's *Invisible Pyramid*.

27 'Theory and Practice: Their Relation to Human Needs' 361

28 Examples are Roszak *Where the Wasteland Ends*, Shepard *The Tender Carnivore*, Ferkiss *The Future of Technological Civilization*.

Works cited

BACON, FRANCIS *The Philosophy of Francis Bacon* edited and translated by B. Farrington. Chicago, University of Chicago Press 1966

- *The Works of Francis Bacon* 7 volumes, edited by J. Spedding, R.L. Ellis, and D.D. Heath. London, Longman, Green 1857-74

BARAN, PAUL and PAUL M. SWEEZY *Monopoly Capital* New York, Monthly Review Press 1966

BARNETT, WILLIAM 'The Effects of Consumer Bliss on Welfare Economics' *Journal of Economic Issues* 7 (1973): 29-45

BAUDRILLARD, JEAN *The Mirror of Production* translated by Mark Poster. St Louis, Mo., Telos Press 1975

- *Pour une critique de l'économie politique du signe* Paris, Gallimard 1972
- *La Société de consommation* Paris, SGPP 1970
- *Le Système des objets* Paris, Gallimard 1968

BAY, CHRISTIAN 'Needs, Wants, and Political Legitimacy' *Canadian Journal of Political Science* 1 (1968): 241-60

BECKER, GARY 'A Theory of Marriage: Part I' *Journal of Political Economy* 81 (1973): 813-46

BERNARD-BÉCHARIES, J.F. *La Choix de consommation: Rationalité et réalité du comportement du consommateur* Paris, Éditions Eyrolles 1970

BOOKCHIN, MURRAY *Post-Scarcity Anarchism* Berkeley, Cal., Ramparts Press 1971

BOULDING, KENNETH *Beyond Economics: Essays on Society, Religion and Ethics* Ann Arbor, University of Michigan Press 1968

BOWLEY, MARIAN *Studies in the History of Economic Theory before 1870* London, Macmillan 1973

BRADBURN, NORMAN and DAVID CAPLOVITZ *Reports on Happiness* Chicago, Aldine 1965

BRAYBROOKE, DAVID 'Let Needs Diminish that Preferences May Prosper' in N. Rescher (ed) *Studies in Moral Philosophy* Oxford, Basil Blackwell 1968

BRUBAKER, STERLING *To Live on Earth: Man and his Environment in Perspective* Baltimore, Johns Hopkins University Press 1972

BRUNNER, JOHN *The Sheep Look Up* New York, Ballantine Books 1972

CAPLOVITZ, DAVID *The Poor Pay More* New York, The Free Press 1963

CARPENTIER, J. 'Organisational Techniques and the Humanisation of Work' *International Labour Review* 110, no. 2 (1974): 93-116

COLE, H.S.D. et al *Thinking about the Future* London, Chatto and Windus 1973

COMMONER, BARRY *The Closing Circle* New York, Bantam 1972

DALTON, GEORGE *Economic Anthropology and Development* New York, Basic Books 1971

- (ed) *Tribal and Peasant Economics* Garden City, NY, Natural History Press 1967

DALY, HERMAN (ed) *Toward a Steady-State Economy* San Francisco, W.H. Freeman 1973

DARLING, F. FRASER and JOHN P. MILTON (eds) *Future Environments of North America* Garden City, NY, Natural History Press 1966

DATOR, JIM 'Neither There nor Then: A Eutopian Alternative to the Development Model of Future Society' in *Human Futures* London, IPC Science and Technology Press Ltd 1974

DEBORD, GUY *Society of the Spectacle* Detroit, Black & Red 1970

DE GRAZIA, SEBASTIAN *Of Time, Work, and Leisure* New York, Doubleday 1964

DICKSON, DAVID *Alternative Technology and the Politics of Technical Change* London, Fontana 1974

DUBOS, RENÉ *So Human an Animal* New York, Scribner's 1968

DUMONT, FERNAND *La Dialectique de l'objet économique* Paris, Éditions Anthropos 1970

DURKHEIM, ÉMILE *Suicide* translated by J. Spaulding and G. Simpson. Glencoe, The Free Press 1951

EHRLICH, PAUL and ANNE EHRLICH *Population, Resources, Environment* 2nd edition. San Francisco, W.H. Freeman 1972

- ANNE EHRLICH and JOHN HOLDREN *Human Ecology* San

Francisco, W.H. Freeman 1973
EISELEY, LOREN *The Invisible Pyramid* New York, Scribner's 1970
ELKIN, LORNE 'Commodity Consumption and Human Need' unpublished paper, Department of Psychology, University of Regina 1970
ENGELBOURG, SAUL 'Insatiability, a Problem for Utopia?' *American Journal of Economic Studies* 22 (1963): 129-40
ETZIONI, AMITAI *The Active Society* New York, The Free Press 1968
EWEN, STUART 'Advertising as Social Production' in H. Reid (ed) *Up the Mainstream* New York, David McKay Co. 1974
FALK, RICHARD *This Endangered Planet: Prospects and Proposals for Human Survival* New York, Vintage Books 1972
FARVAR, M.T. and JOHN P. MILTON (eds) *The Careless Technology: Ecology and International Development* Garden City, NY, Natural History Press 1972
FERKISS, VICTOR *The Future of Technological Civilization* New York, Braziller 1974
FOURIER, CHARLES *Design for Utopia* New York, Schocken Books 1971
FRITSCH, ALBERT *The Contrasumers* New York, Praeger 1974
FROMM, ERICH *The Sane Society* New York, Rinehart 1962
GAEDECKE, R.M. and W.W. ETCHESON (eds) *Consumerism: Viewpoints from Business, Government, and the Public Interest* San Francisco, Canfield Press 1972
GALBRAITH, J.K. *The Affluent Society* Boston, Houghton Mifflin 1958
GEORGESCU-ROEGEN, NICHOLAS *Analytical Economics: Issues and Problems* Cambridge, Mass., Harvard University Press 1967
GINTIS, HERBERT 'Consumer Behavior and the Concept of Sovereignty: Explanations of Social Decay' *American Economic Review* 62 (1972): 267-78
– 'A Radical Analysis of Welfare Economics' *Quarterly Journal of Economics* 86 (1972): 572-99
GIRADIN, JEAN-CLAUDE 'Towards a Politics of Signs: Reading Baudrillard' *Telos* no. 20 (Summer 1974): 127-37
GODELIER, MAURICE *Rationality and Irrationality in Economics* translated by B. Pearce. London, New Left Books 1972
GOODMAN, PAUL and PERCIVAL GOODMAN *Communitas* New York, Vintage 1960
GORZ, ANDRÉ *Socialism and Revolution* translated by N. Denny.

New York, Doubleday 1973
GOULET, DENIS *The Cruel Choice: A New Concept in the Theory of Development* New York, Athenaeum 1971
HARRIS, MARVIN *Culture, Man, and Nature* New York, Crowell 1971
HAUG, W.F. *Kritik der Warenästhetik* Frankfurt, Suhrkamp 1971
HEILBRONER, ROBERT *An Inquiry into the Human Prospect* New York, Norton 1974
- *The Making of Economic Society* 4th edition. Englewood Cliffs, NJ, Prentice-Hall 1972
HELLER, AGNES 'Theory and Practice: Their Relation to Human Needs' *Social Praxis* 1 (1973): 359-73
HENRY, JULES *Culture against Man* New York, Random House 1963
HERSKOVITS, MELVILLE *Economic Anthropology* New York, Knopf 1952
HETTICH, WALTER 'Consumption Benefits from Education' in Sylvia Ostry (ed) *Canadian Higher Education in the Seventies* Ottawa, Information Canada 1972
HIRSCHMAN, ALBERT 'An Alternative Explanation of Contemporary Harriedness' *Quarterly Journal of Economics* 87 (1973): 634-7
- *Exit, Voice, and Loyalty* Cambridge, Mass., Harvard University Press 1970
HOLLIS, MARTIN and EDWARD NELL *Rational Economic Man* London, Cambridge University Press 1975
HORKHEIMER, MAX and T.W. ADORNO *Dialectic of Enlightenment* translated by John Cumming. New York, Herder and Herder 1972
HUDSON, W.D. *Modern Moral Philosophy* New York, Doubleday 1970
ILLICH, IVAN *Energy and Equity* New York, Harper & Row 1974
- *Tools for Conviviality* New York, Harper & Row 1973
JAMES, WILLIAM *Memories and Studies* London, Longmans 1912
JEVONS, W.S. *The Theory of Political Economy* Harmondsworth, Penguin 1970
KATONA, GEORGE *The Mass Consumption Society* New York, McGraw-Hill 1964
- 'What is Consumer Psychology?' *American Psychologist* 22 (1967): 219-26
- B. STRUMPEL, and E. ZAHN *Aspirations and Affluence* New York, McGraw-Hill 1971
KAUDER, EMIL *A History of Marginal Utility Theory* Princeton, Princeton University Press 1965
KNEESE, ALLEN, ROBERT AYRES, and RALPH D'ARGE *Economics and*

the Environment Washington, DC, Resources for the Future 1970
KRIEGER, MARTIN 'What's Wrong with Plastic Trees?' *Science* 179 (2 February 1973): 446-55
KROPOTKIN, PETR *The Essential Kropotkin* edited by E. Capouya and K. Tompkins. New York, Liveright 1975
LANCASTER, KELVIN 'Change and Innovation in the Technology of Consumption' *American Economic Review* 56 *Papers and Proceedings* (May 1966): 14-23
- *Consumer Demand: A New Approach* New York, Columbia University Press 1971
LANDSBERG, HANS, LEONARD FISCHMAN, and JOSEPH FISHER *Resources in America's Future: Patterns of Requirements and Availabilities, 1960-2000* Baltimore, Johns Hopkins University Press 1963
LEE, DOROTHY *Freedom and Culture* Englewood Cliffs, NJ, Prentice-Hall 1959
LEISS, WILLIAM 'Critical Theory and its Future' *Political Theory* 2 (1974): 330-49
- *The Domination of Nature* New York, Braziller, 1972. Paperback edition with new preface, Boston, Beacon Press 1974
- 'The False Imperatives of Technology' in David Shugarman (ed) *Thinking about Change* Toronto, University of Toronto Press 1974
- 'The Imperialism of Human Needs' *The North American Review* 259, no. 4 (1974): 27-34
LÉVI-STRAUSS, CLAUDE *The Raw and the Cooked* translated by J. and D. Weightman. New York, Harper & Row 1969
LINDER, STAFFAN *The Harried Leisure Class* New York, Columbia University Press 1970
LIPSEY, RICHARD, G. SPARKS, and P. STEINER *Economics* 4th edition. New York, Harper & Row 1973
LIVINGSTON, JOHN *One Cosmic Instant: A Natural History of Human Arrogance* Toronto, McClelland and Stewart 1973
LUKÁCS, GEORG *History and Class Consciousness* Cambridge, Mass., MIT Press 1971
MACPHERSON, C.B. *Democratic Theory: Essays in Retrieval* Oxford, Clarendon Press 1973
- *The Political Theory of Possessive Individualism* New York, Oxford University Press 1962
MALINOWSKI, BRONISLAW 'Culture as a Determinant of Behavior' in *Factors Determining Human Behavior* Cambridge, Mass., Harvard

University Press 1937
MANDELL, LEWIS et al *Surveys of Consumers, 1971-72* Ann Arbor, Institute for Social Research 1973
MARCUSE, HERBERT *Counterrevolution and Revolt* Boston, Beacon Press 1972
- *An Essay on Liberation* Boston, Beacon Press 1969
- *One-Dimensional Man* Boston, Beacon Press 1964
MARRIS, ROBIN *The Economic Theory of 'Managerial' Capitalism* Glencoe, The Free Press 1964
MARSHALL, ALFRED *Principles of Economics* 2 volumes, 9th edition. New York, Macmillan 1961
MARX, KARL *Capital* 3 volumes. Moscow, Foreign Languages Publishing House 1961
- *Contribution to the Critique of Political Economy* translated by S.W. Ryazanskaya. Moscow, Progress Publishers 1970
- *Grundrisse: Introduction to the Critique of Political Economy* translated by M. Nicolaus. Harmondsworth, Penguin 1973
MASLOW, ABRAHAM 'A Theory of Human Motivation' *Psychological Review* 50 (1943): 370-96
- *Toward a Psychology of Being* Princeton, NJ, D. Van Nostrand 1962
Material Needs and the Environment Final Report of the National Commission on Materials Policy. Washington, DC, U.S. Government Printing Office 1973
MEADOWS, D.H. et al *The Limits to Growth* New York, New American Library 1972
MENGE, WOLFGANG *Der Verkaufte Käufer* Vienna, Verlag F. Molden 1971
MILL, JOHN STUART *The Principles of Political Economy* 2 volumes, edited by J.M. Robson. Toronto, University of Toronto Press 1965
MILLER, ROGER *Economics Today* San Francisco, Canfield Press 1973
MISHAN, E.J. *The Costs of Economic Growth* New York, Praeger 1967
- *Making the World Safe for Pornography and Other Intellectual Fashions* London, Alcove Press 1973
MOORE, STANLEY 'Marx and the State of Nature' *Journal of the History of Philosophy* 5 (1967): 133-48
- 'The Metaphysical Argument in Marx's Labour Theory of Value' in *Cahiers de L'Institut de Science Économique Appliquée*

supplement no. 140 (*Études de Marxologie* no. 7). Paris, ISEA 1963
MORRIS, WILLIAM *News from Nowhere* New York, Monthly Review Press 1966
MOSCOVICI, SERGE *La Société contre nature* Paris, Union Générale d'Éditions 1972
- and F. COLUMELLI 'Contributions à l'étude des aspirations économiques des travailleurs' *Bulletin du Centre d'Études et Recherches Psychotechniques* 6 (1957): 405-22
PARKER, RICHARD *The Myth of the Middle Class* New York, Harper and Row 1972
PARSONS, TALCOTT *The Structure of Social Action* New York, The Free Press 1961
PASSMORE, JOHN *Man's Responsibility for Nature* London, Duckworth 1974
PERLOFF, ROBERT 'Consumer Analysis' in *Annual Review of Psychology* 19 (1968). Palo Alto, Calif., Annual Reviews, Inc. 1968
PIMENTEL, DAVID et al 'Food Production and the Energy Crisis' *Science* 182 (2 November 1973): 443-9
POLANYI, KARL *The Great Transformation* Boston, Beacon Press 1957
RESCHER, NICHOLAS *Welfare* Pittsburgh, University of Pittsburgh Press 1972
RIVERS, PATRICK *The Restless Generation* London, Davis-Poynter Ltd. 1972
RODMAN, JOHN 'The Dolphin Papers' *The North American Review* 259, no. 1 (1974): 13-26
- 'The Ecology of Openness' in D. Germino and K. von Beyme (eds) *The Open Society in Theory and Practice* The Hague, Nijhoff 1974
ROSZAK, THEODORE *Where the Wasteland Ends* New York, Doubleday 1972
RUSKIN, JOHN *Unto this Last, Munera Pulveris* London, Oxford University Press 1923
SAHLINS, MARSHALL *Stone Age Economics* Chicago, Aldine 1972
SCHERF, GERHARD 'Consumer Education as a Means of Alleviating Dissatisfaction' *The Journal of Consumer Affairs* 8 (1974): 61-75
SCHNEIDER, MICHAEL 'Neurosis and Class Struggle: Toward a Pathology of Capitalist Commodity Society' *New German Critique* 1, no. 3 (1974): 109-26
SCHUMACHER, E.F. *Small is Beautiful: Economics as if People*

Mattered New York, Harper and Row 1973

SCHUMPETER, JOSEPH *A History of Economic Analysis* New York, Oxford University Press 1954

SCITOVSKY, TIBOR *Papers on Welfare and Growth* London, Allen and Unwin 1964

SHEPARD, PAUL *The Tender Carnivore and the Sacred Game* New York, Scribner's 1973

SILVERMAN, MILTON and P.R. LEE *Pills, Profits and Politics* Berkeley, University of California Press 1974

SLATER, PHILIP *The Pursuit of Loneliness* Boston, Beacon Press 1970

SMITH, ADAM *An Inquiry into the Nature and Causes of the Wealth of Nations* edited by E. Cannan. New York, Modern Library 1937

STANLEY, MANFRED 'Nature, Culture, and Scarcity' *American Sociological Review* 33 (1968): 855-70

STELLMAN, J.M. and S.M. DAUM *Work is Dangerous to your Health* New York, Vintage Books 1973

STONE, CHRISTOPHER *Should Trees have Standing? Toward Legal Rights for Natural Objects* Los Altos, Calif., William Kaufmann, Inc. 1974

TAWNEY, R.H. *The Acquisitive Society* London, G. Bell and Sons 1952

THOMAS, W.L. (ed) *Man's Role in Changing the Face of the Earth* Chicago, University of Chicago Press 1956

TRIBE, LAURENCE 'Technology Assessment and the Fourth Discontinuity: The Limits of Instrumental Rationality' *Southern California Law Review* 46 (1973): 617-60

– 'Ways not to think about Plastic Trees: New Foundations for Environmental Law' *The Yale Law Journal* 38 (1974): 1315-48

TWEDT, D.W. 'Consumer Psychology' in *Annual Review of Psychology* 16 (1965). Palo Alto, Calif., Annual Reviews, Inc. 1965

VAYDA, ANDREW (ed) *Environment and Cultural Behavior: Ecological Studies in Cultural Anthropology* Garden City, NY, Natural History Press 1969

VEBLEN, THORSTEIN *The Theory of the Leisure Class* New York, New American Library 1953

VICTOR, PETER *Pollution: Economy and Environment* Toronto, University of Toronto Press 1972

WEISSKOPF, WALTER *Alienation and Economics* New York, Dutton 1971

- *The Psychology of Economics* Chicago, University of Chicago Press 1955

WOLIN, SHELDON 'Looking for "Reality" ' *The New York Review of Books* 22, no. 1 (6 February 1975): 15-20

ZAMYATIN, YEVGENY *We* translated by M. Ginsburg. New York, Bantam 1972

ZWERDLING, DANIEL 'Death for Dinner' *The New York Review of Books* 21, no. 2 (21 February 1974): 22-4

Acknowledgments

Many years ago some friends and I, fellow graduate students at a well-regarded institution of higher learning, happened upon one of our professors hard at work in the drawers of the library's card catalog. We were shocked by this waste of expensive talent, and not a little annoyed at the apparent threat to our livelihood, since bibliographical research was one of the more exalted functions normally reserved for graduate assistants. In response to our expressions of concern the professor apologized for his unusual position and explained the delicate nature of his mission. He was seeking to identify all the recognized authorities who had published in the same field as his own and who therefore might be called upon to review his forthcoming book; his objective, he announced, was to insert in his book a laudatory footnote to their works in the hope of compromising them.

Since in this essay I have poached rather wantonly in fields lying outside the borders of my own professional domains, I decided that it would be pointless to emulate this author's precautionary measures. Instead, I have relied heavily on the assistance of friends and colleagues to rescue me from the many unsuspected pitfalls that I encountered in exploring the huge and treacherous terrain that has been surveyed here. I am especially grateful for Sandra Monteath's page-by-page editorial commentary and Peter Victor's extensive marginal notations on the first draft, and for Michael Neumann's incisive critique of my arguments. Others who read part or all of the manuscript at some stage and who offered criticisms and much-needed encouragement are: Judy Adler, Christian Bay, Andrew Feenberg, Cliff Hooker, Peter Leiss, Bill Livant, John Livingston, C.B. Macpherson, Sue Mansfield, Herbert Marcuse, Chuck and Dian Marino, Volker Meja, Peter Penz, Ron Perrin, Jo Roberts, John Rodman, Abe Rotstein, Morton Schoolman, Edwin Seaver, Ricky Sherover, and Mark Taylor. I have only myself to blame for failing to respond adequately in revising the manuscript to some of the objections which they raised.

I began exploring the subject of this essay in a graduate seminar at the University of Regina in 1969-70. During the time I spent in Regina, Bill Livant (who has been a most perceptive and loyal critic of my work), Joe Roberts, Dallas Smythe, and many others created an especially supportive atmosphere for our shared activities. In the Faculty of Environmental Studies at York University I have enjoyed a close association with Grahame Beakhust, John Livingston, and John Page; I have learned much from them and other colleagues there that changed my perspectives on the tasks of contemporary social theory. To Dean Gerald Carrothers I am indebted for indispensable personal and administrative support during the period of research and writing. Members of York University's staff – Lisa Chapnick, Sheila Herbert, Lillian Kindree, Phyllis Kroeker, and Joan Shields, together with Sylvia Zingrone (who co-ordinated the arrangements) – were most helpful in the preparation of the manuscript. The enthusiastic co-operation of Arlin Hackman and the aid of other graduate students at York – Judy Besse, Peter Meiksins, Christopher Olive, Pat Patterson, and Diane Willemse – was essential to bring this project to completion. Finally, I wish to thank Ray Jackson and Arthur Cordell for encouraging me to reformulate parts of this study in the context of the Conserver Society Project undertaken by the Science Council of Canada.

Edwin Seaver, Armand Petitjean, and Roberta Fitzsimmons made the relationship of editor and author a most pleasurable experience for me during the publication of my first book. In addition to those already mentioned, many other friends have been very generous in their expressions of support over the years: Mildred Bakan, Steve Berkowitz, John Burke, Robert Cohen, Sue and John Drysdale, Fred and Lou-Jean Fleron, Steve Levine, Sandy and Betsy Lockhart, Stanley Moore, David Ober, Paul Piccone, Herbert Reid, David Shugarman, Kurt Wolff, Ellen and Neal Wood, and Irving Zeitlin.

My editor at the University of Toronto Press, Rik Davidson, has guided this manuscript along the way to publication with great care and skill. The reader for the press was both sympathetic and helpful in his critical remarks, and Margaret Parker's editing of the manuscript made a much better book out of it than it otherwise would have been.

The Canada Council and the Louis M. Rabinowitz Foundation provided small grants in aid of research at an early stage of the work on this project.

No conventional form of acknowledgment could convey what my association with Herbert Marcuse has meant to me. Since as a teacher he seeks no disciples, the imprint of his influence does not necessarily show itself in his students' work. His philosophical interpretations and arguments always challenged us; but I recall best the spontaneous bond of respect and affection which began to develop fifteen years ago, when quite by accident I enrolled in one of his courses, and which has steadily grown stronger during the intervening years. The unique combination of learning and commitment, of insight and modesty, of seriousness and wit that he brings to all his activities and relationships has been for me a source of enduring delight and inspiration.

In the dedication of this book I have sought to indicate my gratitude for what I cannot express in writing to those who have been closest to me.